Physics: The Fabric of Reality

PHYSICS:
The Fabric of Reality

S. K. KIM   MACALESTER COLLEGE

Macmillan Publishing Co., Inc.
NEW YORK
Collier Macmillan Publishers
LONDON

Macmillan Publishing Co., Inc.
866 Third Avenue, New York, New York 10022

Collier-Macmillan Canada, Ltd.

Library of Congress Cataloging in Publication Data

Kim, Sung Kyu (date)
  Physics, the fabric of reality.

  Includes bibliographies.
  1. Physics. I. Title.
QC21.2.K52   530   73-14015
ISBN 0-02-363770-6

Printing: 1 2 3 4 5 6 7 8   Year: 5 6 7 8 9 0

To Robert Eliot and Jennifer Ruth

Preface

> One thing I have learned in a long life: that all our science, measured against reality, is primitive and childlike—and yet it is the most precious thing we have.[1]
>
> ALBERT EINSTEIN

New insights are precious. A particular cause for marvel is the physical insights that have been achieved during the twentieth century. They constitute a truly profound intellectual revolution. This revolution, however, has not been fully appreciated by all, for the modern world of intellect still consists of "two cultures."

The physics community in higher education can do much to bridge the intercultural gap. Recent efforts to reach the students outside the scientific disciplines are, clearly, a positive development in physics education. Of particular significance is a proliferation of courses with a primary emphasis on modern physics. They are, by and large, receiving enthusiastic response from the students.

The present work seeks to share the excitement of modern physics with nonscience students. The book begins with the theory of relativity. There are compelling pedagogical advantages in such an approach. No topic in physics stirs the imagination of liberal arts students as quickly and broadly as relativity. It thus helps create an exciting climate in which to teach and to learn. Furthermore, relativistic concepts can be discussed, without sacrifice of logical rigor, in the language of the students. This fact

[1] B. Hoffman and H. Dukas, *Albert Einstein: Creator and Rebel* (New York: Viking Press, 1972).

vii

tends to involve the students more actively in the thought process, and thus enhances the quality of their experience.

The chapter on Newton's theory of gravitation appears here primarily as a way of putting Einstein's work in perspective. The chapters on electric and magnetic fields lay groundwork for the electromagnetic theory of light. The field ideas are familiarized by application to accessible and interesting examples in the students' environment. Relativistic insights into electromagnetism are carefully explored.

The discussion of the electromagnetic nature of light and its interaction with matter leads naturally to quantum ideas. Quantum phenomena are, perhaps, no more mysterious than, say, the phenomenon of tides, but the concepts developed to describe them represent radical departures from the familiar. Thus perplexity and fascination abound, and the realization that the quantum reality, like others, is a creation of the human mind leaves on the students a profound impression.

The choice of topics has been influenced by the desire to show the power of rational and imaginative thought. The arrangement of topics has been dictated by the need to maintain a smooth transitional flow of ideas. A topic has frequently been introduced early for its value in a later part of the book, and later chapters often touch on material covered earlier in a different context.

I believe with Heisenberg that the fundamental ideas of physics "can be described in the language of daily life." I have shared them during the past ten years with liberal arts students at Macalester College and at the University of California at Irvine. I think it safe to say that few, if any, other fields of human endeavor have revealed as much of beauty and wonder as physics.

S. K. K.

Acknowledgments

The opportunity to share the excitement of physics with liberal arts students has been a challenge as well as a joy to me. To my colleagues at Macalester College who made it possible I owe a special debt of gratitude: Professors E. N. Strait, J. H. Roberts, R. C. Mikkelson, and R. B. Hastings.

I am indebted to many authors. In particular, I have been enriched, and most directly influenced, by *Space and Time in Special Relativity* by N. D. Mermin, *Lectures on Physics,* Volumes I, II, and III, by R. P. Feynman, *The Conceptual Development of Quantum Mechanics* by M. Jammer, and *Basic Physics* by K. W. Ford. Articles in *Scientific American* and in *Physics Today* have also been sources of information and inspiration.

Without the personal encouragement of Professor K. W. Ford this book would not have been likely to reach completion. I am particularly grateful to him for reading the entire manuscript and offering many helpful suggestions. Professors H. O. Hooper, W. W. Kelly, and D. R. Rutledge reviewed the manuscript in various stages of development, and for their supportive comments I owe them a debt of gratitude. Any errors are certainly my own.

My students, both at Macalester College and at the University of California at Irvine, deserve a special word of thanks. It was their enthusiastic response to this unconventional presentation of physics that prompted me to embark on the present project. In particular, I wish to thank Magda Krance for helping with the illustrations and Wendy Knox for editing and typing the earlier chapters.

My acknowledgments of indebtedness would certainly be incomplete without recognition of my wife's major role. In addition to serving as a source of constant encouragement and moral support, she spent countless hours in direct help in the preparation of the manuscript.

S. K. K.

Acknowledgments

Contents

Contents

Physics: The Fabric of Reality

1

Motion: Absolute or Relative?

Man has always been curious about the universe. The heavens have, through the ages, been a source of boundless fascination, and their rich diversity has been a cause for marvel and wonder. As patterns emerged amidst the diverse motions of heavenly bodies, thinkers probed deeply into the mysteries of the universe and pondered man's and the earth's place in its seeming infinitude.

Aristotelian View of Motion

Foremost among ancient thinkers was Aristotle (384–322 B.C.). An Athenian physicist, logician, and philosopher, Aristotle dominated the intellectual climate of ancient Europe. His was perhaps the first model of the cosmos to be proposed in the intellectual tradition of the West. The Aristotelian cosmology was based on the vision of a closed universe —a sphere decorated with lesser spheres of celestial bodies in their incessant circular motions about the center of the universe. The same cosmology ascribed a privileged status to earth and placed it at the immovable center of the universe. Existing in a state of absolute rest, the earth was thus set apart from the rest of the universe.

As a proof of the earth's state of absolute rest, Aristotle proffered an argument that if the earth were in a state of motion, an object thrown

FIGURE 1–1.   The Galilean pilot playing an Aristotelian game.

vertically upward on the surface would not return to its initial position. He then pointed to the observation that an object so tossed does, indeed, return to the same position. He thus sought to establish absolute rest as a natural state for earth and ascribed absolute motion to all heavenly bodies to be their natural state of being. It was, in fact, this absolute character of rest and motion that formed a foundation for Aristotelian geocentric cosmology as well as for Aristotelian physics.

Galileo's Challenge

It was Galileo Galilei (1564–1642) who successfully challenged the tradition of Aristotelian physics. The publication of Galileo's *Two New Sciences,* in 1638, was instrumental in exposing the falsity of Aristotelian claims about the nature of motion. To Galileo the observation and the accompanying argument put forth by Aristotle in support of the earth's state of rest constituted no proof at all. He proffered a counterargument. Imagine a boat sailing on a calm sea along a straight course at a constant speed. Suppose the pilot throws an object vertically upward to establish, as claimed by Aristotelian physics, the boat's state of motion. The object

FIGURE 1–2. The Galilean
boat as observed from the
shore.

would return to the initial position in the boat, whence the pilot, if he
were an Aristotelian physicist, would ascribe a state of rest to the
boat.

Suppose now that the pilot's little game is observed from the shore.
The Aristotelian method would lead the shore-bound observer to ascribe
a state of rest to the Galilean boat, but in actuality he would ascribe a
state of motion to the boat. Galileo thus pointed out that the absolute character of rest and motion—and, more importantly, the earth's
state of absolute rest—could not be established by the Aristotelian
method.

Is there, then, another way of establishing a state of absolute rest? Or
is it that the boat's state of motion, as perceived by the shore-bound
observer, must be perceived as a state of rest by the Galilean pilot? The
latter alternative would undermine the very foundation of Aristotelian
physics, as it stands contrary to the absolute character of rest and
motion.

Motion: Absolute or Relative?

Frames of Reference

Viewing it in isolation, the Galilean pilot would, indeed, ascribe a state of rest to the boat. With respect to the shore, however, he would view the boat to be clearly in motion. But viewed either way, everything in the boat remains exactly the same. The distinction between the states of rest and motion thus recedes, and the notion of motion—at least, of the Galilean type—becomes quite ambiguous apart from a reference point. In consequence, a new concept, called a *frame of reference*—or a coordinate system—emerges and becomes an integral part of motion.

The concept of a coordinate system is a familiar one. The index to a map, for example, is based on the concept of a rectangular coordinate system, where the position of a town is usually designated by an Arabic numeral and an English letter.

A commonly used coordinate system, called the Cartesian system, consists of three mutually perpendicular reference lines, all originating at a common point. The reference lines are referred to as *x, y,* and *z* axes. The position of an object is, then, specified by a set of ordered numbers (x,y,z), which give the perpendicular distances to the position of the object from the corresponding axes. For example, a set of ordered

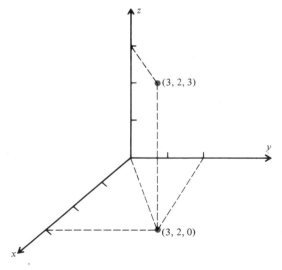

FIGURE 1–3. A rectangular coordinate system.

numbers (3, 2, 3), in appropriate units of length, specifies a unique position in space relative to the origin of a definite coordinate system.

Uniform Motion

In addition to position, the concept of velocity requires a coordinate system. Velocity is defined as a ratio of the distance traversed by a moving object to the time required:

$$\text{velocity} = \frac{\text{distance traversed}}{\text{time elapsed}}$$

(Although velocity and speed have distinct meanings in physics, we shall use the terms interchangeably here, because confusion is unlikely.)

The simplest type of velocity is a constant velocity. An object moving at a constant velocity traverses equal distances in equal intervals of time. Motion at a constant velocity along a straight line constitutes a special type of motion and is referred to as *uniform motion*.

A unique feature of uniform motion is that it is indistinguishable in its effects from nonmotion. Consider, as an example of uniform motion, the case of the Galilean boat. The pilot would assert that he feels no effects of the motion which the shore-bound observer ascribes to his boat. (The idealized sea, of course, is perfectly smooth.) In fact, he could claim that his boat is not in motion at all, but rather that it is the water and the shore together that pass under the boat and away from him. The turbulence one often associates with certain types of motion would, at least, not be a part of his experience.

Or view uniform motion on a grander scale. An observer on earth, named Pongo, looks out his window and catches a fleeting glimpse of a planet streaking through the universe.

"Look," he says, drinking a cup of coffee. "How could anyone live on a planet moving so fast?" However, unbeknownst to Pongo, on that very planet sits another observer, Ognop, placidly drinking a cup of eeffoc, who remarks to his student, "Ydnew! Did you see that universe streak by? They must have been going a million miles per ruoh!" Neither of them is disturbed, neither spills his drink, and neither of them experiences any effects of motion. Why? Because the effects of uniform motion are indistinguishable from the effects of nonmotion.

Motion: Absolute or Relative?_____

5

FIGURE 1–4. The pilot's claim that motion is all around him.

In fact, there is no experiment that would distinguish the state of rest from the state of uniform motion. There is no inherent difference between uniform motion and nonmotion. What appears as a state of uniform motion to one observer would appear as a state of rest to another observer. The captain of a jetliner, for example, is not in motion when observed by a seated passenger, but to ground-based observers he is in as rapid motion as the plane itself. Uniform motion, therefore, is a relative concept.

Galilean Principle of Relativity

An observer who has no physical way of determining, even in principle, whether he is at rest or in uniform motion is called an *inertial* observer. An inertial observer would observe that, in the absence of external disturbances, an object at rest remains at rest and an object in uniform motion persists in uniform motion. First abstracted from observation by Galileo, this fact constitutes a law of nature and is referred to as the law of inertia. Any reference system in which the law of inertia holds valid is thus an inertial frame of reference. There are, of course, non-inertial reference frames. An airplane encountering air turbulence is an example of a noninertial frame of reference; a coffee cup on such a plane would, of course, not remain stationary.

The concept of the inertial frame is useful, because if motion is relative, all inertial frames of reference must be equivalent. This equivalence implies that no physical phenomenon would be a unique property of a preferred frame of reference. Insofar as physical phenomena are concerned, all inertial observers thus enjoy a perfect form of democracy.

It was Galileo who recognized the profound significance of the equivalence of all inertial frames. He envisioned this equivalence to pervade all motion-related phenomena, so that a physical law was an acceptable law only when it was a valid law in all frames of reference. He thus gave rise to a physical principle, which is now referred to as the *Galilean principle of relativity*. This principle of relativity requires that the laws of mechanics be formulated so as to be the same to all inertial frames of reference. Physical laws are thus accorded a quality of invariance.

In short, it is clear that a fatal blow has been delivered to the Aristotelian notion of absolute motion. It is also clear that the relative nature of motion, amply substantiated by experience, forms the foundation for Galilean physics. Nevertheless, we somehow find the notion of absolute motion difficult to part with, and wonder, almost instinctively, if there might not still exist an absolute frame of reference somewhere in this vast universe of ours to resurrect absolute motion to a post-Galilean glory.

QUESTIONS

1–1. Suppose the pilot in the Galilean boat tosses an object vertically upward. How would the path of the object appear to the observer on the shore, as it rises and falls?

1–2. If the shore-bound observer tosses an object vertically upward, how would the path of the object appear to the pilot of the Galilean boat?

1–3. Could the Galilean argument disprove the earth's unique position at the center of the universe?

Motion: Absolute or Relative?_____

1–4. It is said of the great mathematician John von Neumann that after an auto accident at Princeton he emerged from his wrecked car to explain, "The trees on the right were passing me in orderly fashion at 60 miles per hour. Suddenly one of them stepped out in my path," What arguments could an attorney employ in defense of von Neumann's statement?

1–5. Design a simple thought experiment to establish that one is an inertial observer. Argue that all observers in uniform motion relative to an inertial observer are also inertial observers.

1–6. One could arrive at the Galilean principle of relativity by observing that all the existing laws of mechanics have the same mathematical form in all inertial frames of reference. Or one could start with the Galilean principle of relativity, and require that the laws of mechanics be formulated so as to have the same mathematical form in all inertial frames. Is there a fundamental difference in the two viewpoints? Explain.

1–7. If an object, in the absence of all external influences, is observed to be at rest, what possible states of motion would another inertial observer ascribe to the object? State clearly the principles you invoke in reaching your answer.

SUGGESTIONS FOR FURTHER READING

ARONS, A. *Development of Concepts of Physics*. Reading, Mass.: Addison-Wesley Publishing Co., Inc., 1965. (Chapter 2.)

MARION, J. B. *A Universe of Physics*. New York: John Wiley & Sons, Inc., 1970. (Sections 2.3–2.6.)

MERMIN, N. D. *Space and Time in Special Relativity*. New York: McGraw-Hill Book Company, 1968. (Chapter 1.)

SMITH, J. H. *Introduction to Special Relativity*. New York: W. A. Benjamin, Inc., 1965. (Chapter 1.)

2

Ether as an Absolute Reference Frame

Among the early admirers of Galileo and his view of science, Sir Isaac Newton (1642–1727) is perhaps the best known. Newton, like Galileo before him, was intrigued by the nature of motion. As he meditated on the observed facts about the nature of motion, the arbitrary distinction between terrestrial and celestial motions receded. The new insight he achieved was that one and the same set of laws describe all motions, those of earth-bound objects as well as those of heavenly bodies. These laws, despite their applicability to the entire universe, were, however, valid only with respect to the inertial frames of reference. Consequently, no physical phenomenon described by Newton's laws of motion would establish an absolute frame of reference. The notion of absolute motion was thus absent in the Newtonian world view.

A Brief Remark About Electromagnetic Waves

There, however, exist other sets of phenomena in the universe that are not described by Newton's laws of motion. Of those, the electromagnetic phenomena are a more familiar example. The science of electricity and magnetism developed mostly during the nineteenth century through the experimental discoveries of physicists such as Oersted, Ampere, and

Faraday. By the middle of the century the science of electricity and magnetism had matured in stature to rival the science of optics, which, as a branch of physics dealing with the behavior of light, had achieved considerable sophistication by the beginning of the century.

In the 1860's James Clerk Maxwell (1831–1879) discovered that the then-known laws of electricity and magnetism could be summarized in a set of mathematical equations. These laws, as they then existed, however, were mutually inconsistent. As he sought to remove the inconsistency, he discovered that he had to introduce a hypothetical term to his equations. But its addition meant that he was proposing the existence of a new phenomenon that his experimentalist contemporaries had failed to observe. Rather than awaiting a confirmation in the laboratory, he dared to believe that nature must be consistent, and on such faith, forged ahead with his work. The result was a set of mathematical equations, which were distinguished by their utter self-consistency as well as structural elegance. These are now referred to as Maxwell's equations, and they describe the totality of macroscopic electromagnetic phenomena.

The new phenomenon that Maxwell hypothesized in order to make his equations self-consistent led to the prediction of a particular type of waves called *electromagnetic waves*. When the speed of the propagation of these waves in empty space was calculated, Maxwell discovered that it was identical to the experimental value of the speed of light in vacuum. On the strength of this identity Maxwell concluded that electromagnetic waves and light are one and the same thing. The hitherto-separate science of optics was thus brought into union with electromagnetism, a feat for which Maxwell's equations are most celebrated.

Electromagnetic Ether

Maxwell's equations, however, were not consistent with the Galilean principle of relativity. This implied that the phenomena described by Maxwell's laws would distinguish the state of motion from the state of rest. The equivalence of all inertial frames of reference, as formulated by Galileo and Newton, were thus no longer valid as a general principle. Maxwell's theory, therefore, revived the notion of the absolute frame of reference and the concept of absolute motion.

The possible existence of an absolute frame of reference becomes most apparent in the case of electromagnetic waves. Maxwell's electromagnetic theory predicts that light is an electromagnetic wave, propagating at 300 million, or 3×10^8, meters per second in empty space. But it is well known that a common property of all familiar waves (such as sound and water waves) is that waves require a medium; for a wave is nothing more than a disturbance in a medium. A pebble tossed to the center of a calm pond, for example, starts a series of circular ripples which move toward the edges. These advancing ripples represent a sequence of steps where the original disturbance affects the medium next to it, each new disturbance repeating the process again.

The speed of propagation of a wave is a characteristic of the medium. The speed of sound in air, for example, is a property of the air itself and has nothing to do with whether the sound has been produced by a human voice or a musical instrument.

It is then reasonable to interpret that the "vacuum" of Maxwell's theory, in which electromagnetic waves are supposed to propagate, cannot represent a true emptiness. For otherwise an empty space would be serving as a medium for electromagnetic waves. Furthermore, a vacuum would possess the unique property that light propagates at a speed of 3×10^8 meters per second with respect to it. But a true vacuum, by definition, cannot have physical properties of any kind. Therefore, it was proposed that the "empty space" of Maxwell's theory be clothed with a physical substance to pervade all space. This hypothetical substance was called *electromagnetic ether*.

Maxwell himself regarded the ether hypothesis as an essential addition to his theory. He wrote:

> The vast interplanetary and interstellar regions will no longer be regarded as waste places in the universe, which the Creator has not seen fit to fill with the symbols of the manifold order of His kingdom. We shall find them to be already full of this wonderful medium . . . It extends unbroken from star to star; and when a molecule of hydrogen vibrates in the Dog-Star, the medium receives the impulse of these vibrations; and after carrying them in its immense bosom for three years, delivers them in due course, regular order, and full tale into the spectroscope . . . at Tulse Hill.

In addition to its all-pervasiveness, the ether had to be frictionless, transparent, incompressible, and of uniform density. If it provided a friction, the earth would slow down in its orbital motion about the sun, as it must float in the sea of ether. That it must be incompressible and uniformly spread throughout the universe is also obvious, for otherwise the velocity of light would be different at different places in space. Finally, if ether were not transparent, stars would be invisible, and darkness would prevail over the universe. Ether was thus a peculiar kind of a substance.

Ether as an Absolute Frame of Reference

An exciting implication of the ether hypothesis was that ether might serve as an absolute frame of reference. Light could then be viewed as propagating at the predicted speed of 3×10^8 meters per second relative to the ether frame of reference.

The ether frame of reference as an absolute frame revived the notion of absolute motion. One could, once again, speak of the absolute velocity of an object—namely, its velocity relative to the ether frame. The earth, if not in a state of absolute rest, would, at least, exist in a state of absolute motion. In fact, all motions of the universe could now be referred to this absolute frame of the omnipresent ether.

The electromagnetic ether was, indeed, a marvelous substance, filling the universe with precision, supporting light in all its journeys, and yet exerting no resistance of any kind. But more important, the ether hypothesis relieved the troubled physicist of the unbearable burden of imagining electromagnetic waves without a physical medium.

But does ether really exist? No matter how pleasing an idea is, the ultimate question of physical reality must be settled in the laboratory. The ether, to be real, had to be experienced. An experiment, therefore, had to be devised and performed.

QUESTIONS

2–1. Professor R. P. Feynman writes, "The most dramatic moments in the development of physics are those in which great syntheses (of knowledge) take place . . . the basis of the success of physical science is mainly that we are *able* to synthesize." Comment on this statement in relation to an intellectual discipline with which you are familiar.

2–2. The laws of electricity and magnetism, up to Maxwell's time, were all discovered experimentally and were quite satisfactory as far as observations were concerned. How can you reconcile this fact with Maxwell's discovery of a mutual inconsistency in the laws?

2–3. What was your initial reaction to the ether hypothesis? List some properties, other than those listed in the text, that ether might possess?

2–4. If the existence of the electromagnetic ether were experimentally established, would the notion of absolute motion have to be introduced to the Newtonian world view also? Explain on philosophical grounds.

2–5. Design a thought experiment to detect the presence of the ether.

SUGGESTIONS FOR FURTHER READING

FORD, K. W. *Basic Physics*. Waltham, Mass.: Blaisdell Publishing Co., 1968. (Section 19.6.)

SMITH, J. H. *Introduction to Special Relativity*. New York: W. A. Benjamin, Inc., 1965. (Chapter 2.)

Ether as an Absolute Reference Frame_____

3

The Michelson–Morley Experiment

Great theories motivate great experiments. Maxwell's electromagnetic theory, published in 1867, was a truly great theory, perhaps the greatest since Newton's theories of motion and gravitation. It achieved a great synthesis of knowledge by uniting the hitherto unrelated fields of optics and electromagnetism. The world, for the first time, was given a glimpse into the basic nature of the mysterious yet marvelous stuff called light.

Maxwell's theory provided the inspiration for two of the greatest experiments in the history of physics. One was the experiment of Heinrich Hertz, in 1887, which demonstrated the validity of the electromagnetic theory by producing and detecting electromagnetic radiation of the type Maxwell had predicted. A precursor of the development of radio waves, the Hertz experiment wrought a revolution in communication.

The other experiment was performed by two American experimentalists, Albert A. Michelson and Edward W. Morley, who also published results in 1887. The experiment was designed to detect ether, the elusive substance that was thought to exist as a medium for light waves.

The Experimental Setup

Like many of the most significant experiments in physics, the Michelson–Morley experiment is conceptually very simple. It essentially seeks to determine the absolute velocity of the earth with respect to the ether.

To start with, let us assume that the earth does move in the sea of ether. Such an assumption is not far-fetched, because the earth revolves about the sun. If the earth and the ether happen to flow in the same

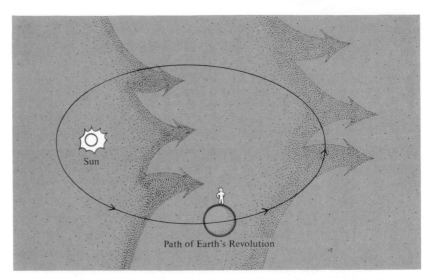

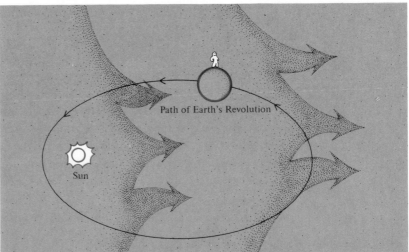

FIGURE 3–1.   The earth in the sea of ether (a) if the ether current and the earth moved in the same direction, (b) then they would certainly be moving in opposite directions six months later.

Physics: The Fabric of Reality_____

direction at any particular time, thus rendering their relative motion un-detectable, they would certainly be flowing in opposite directions six months later. The relative velocity between ether and earth should there-fore be detectable at least some of the time during the year.

A cosmic observer would see the earth floating in the vast sea of ether. For an earth-bound observer, however, it would be more natural to view the ether as sweeping across the earth—across Michelson and Morley and their elaborate apparatus.

This experimental apparatus consisted of a light source, a beam-splitting half-silvered mirror, two reflecting mirrors, and a viewing tele-scope. A typical distance between the beam-splitter and the reflecting mirrors was 120 cm. The reflecting mirrors were placed at the ends of two mutually perpendicular arms of equal length, so that when one arm was parallel to the direction of the ether current, the other would be at right angles to it.

Light from the source was then split into two beams at right angles to each other by the half-silvered mirror at the center. Upon reflection by the mirrors M_1 and M_2, the beams were joined by the half-silvered mirror and sent to the viewing telescope, both beams traversing equal distances.

The conceputual basis for such a setup was the expectation that the light-beam traveling *parallel* to the ether current would require a different interval of time for the completion of a round trip than the beam traveling

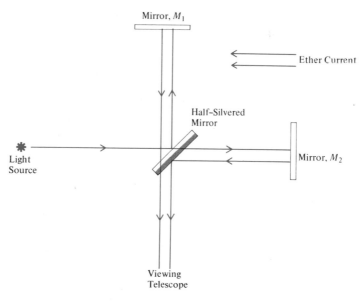

FIGURE 3–2. A schematic of the Michelson–Mor-ley apparatus.

The Michelson–Morley Experiment

across the ether current. More precisely, the time difference would have a unique dependence on the velocity of the ether current. The measurement of just such a time difference was thus at the heart of the experiment.

The Analogy of a Boat and a River

The situation was a little reminiscent of trying to measure the speed of a river current with a motor boat whose speed relative to the current could not be varied. Although no one would measure the speed of a river current in that way, it can be done, and more importantly, the analysis is easy to follow and analogous to the Michelson–Morley experiment.

Put a pilot in our special boat and have him pilot it up a straight river and then straight back down again, traversing equal distances either way. Consider, first, the trip against the current. If the boat's speed with respect to the current is c and the current flows at a velocity v in reference to the bank, then the boat's apparent speed to an observer on the bank is the difference, $c - v$. This difference simply expresses the common-sense observation that a boat bucking a current cannot move as quickly as if there were no current at all. The time it would take for the boat to traverse

FIGURE 3–3. Boat bucking the current.

Physics: The Fabric of Reality

a distance L upstream is then given by dividing the distance by the speed of the boat relative to the river bank. That is,

$$t_{up} = \frac{L}{(c - v)}$$

On the return trip, however, the boat is helped by the current, and the boat would appear to be moving at a higher velocity to an observer on the bank—the apparent velocity being the sum of the current and the boat velocities. The downstream trip, therefore, does not require as much time, as shown by

$$t_{down} = \frac{L}{(c + v)}$$

The total time for a round trip up- and downstream is simply the sum of the corresponding times. That is,

$$t_{up\,+\,down} = t_{up} + t_{down}$$

or

$$t_{up\,+\,down} = \frac{2L}{c\left(1 - \dfrac{v^2}{c^2}\right)}$$

FIGURE 3–4. Boat with the current.

The Michelson–Morley Experiment

Consider now a trip *across* the current. The boat must be directed diagonally into the current in order to travel straight across the river and not be carried downstream. The boat would then be observed crossing the current at a lesser speed, u, relative to the bank, although its speed with respect to the water is c. This apparent speed of the boat is, by the Pythagorean theorem, related to the boat's speed relative to the current and the current speed relative to the bank in the following way:

$$c^2 = u^2 + v^2 \qquad \text{(right triangles)}$$

or

$$u = c\sqrt{1 - \frac{v^2}{c^2}} \qquad \text{(by means of algebra)}$$

The total time required to make a round trip across the current is then given by

$$t_{\text{across}} = t_{\text{away}} + t_{\text{return}} = \frac{2L}{c\sqrt{1 - \dfrac{v^2}{c^2}}}$$

Therefore, the analysis leads to the conclusion that it takes less time to pilot a boat across a river than upstream and back. More precisely,

$$\frac{t_{\text{across}}}{t_{\text{up + down}}} = \sqrt{1 - \frac{v^2}{c^2}}$$

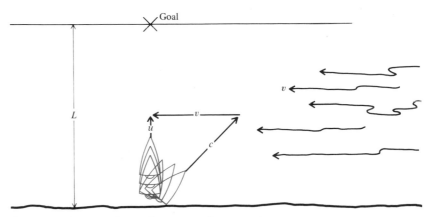

FIGURE 3–5. Boat in motion across the current.

Physics: The Fabric of Reality

The quantity in the radical can never exceed unity, and is always less than unity if the current is flowing at all, that is, $v > 0$.

The Experimental Result

Applying the result of our boat analogy to the Michelson and Morley experiment, we expect that the time of transit for the two mutually perpendicular light beams would be different, and the measurement of this time difference would yield the velocity of the ether current with respect to the earth.

Since they did not know which way the ether current was flowing, Michelson and Morley rotated the apparatus. Possible distortions and vibrations during rotation were eliminated by mounting the apparatus on a massive stone floating in mercury.

The genius of Michelson and Morley lay in their ability to detect to fantastic precision the difference in the transit times of the beams. An ether drift velocity as low as 1 kilometer per second was unambiguously measureable with their apparatus. In view of the fact that the earth revolves about the sun at the considerable speed of 30 kilometers per second, the expectation of a positive, conclusive result was very high.

With the experimental "bugs" out, Michelson and Morley carefully rotated their apparatus. As they looked through their viewing telescope, they expected the light beams to arrive one ahead of the other. They were justifiably confident that the slightest difference in their arrival times would be clearly observable. After all, the present apparatus was a painstaking refinement of a similar experiment performed some six years earlier. The moment came—and a total surprise! Both beams arrived in the telescope at exactly the same time!

Lest the earth should have momentarily been at rest in the sea of ether, the observations were continued in spring and in autumn. If the earth moved in one direction in spring, its orbital motion would be in the opposite direction in autumn. The result, however, was always the same. No matter when the observation was made, the light beams always arrived together.

This ingenious attempt to measure the velocity of the ether had apparently failed. Or could it be that Michelson and Morley did, indeed, succeed in measuring the velocity of the ether, which is simply always zero?

QUESTIONS

3–1. When we say that the earth revolves about the sun, where is our frame of reference?

3–2. If the light beams in the Michelson–Morley experiment traversed a distance of 3 meters, how long would they have been in transit?

Answer: 10^{-8} sec

3–3. Show that

$$t_{across} = \frac{2L}{c\sqrt{1 - \dfrac{v^2}{c^2}}}$$

3–4. Show that

$$t_{across} = \sqrt{1 - \frac{v^2}{c^2}} \times t_{up + down}$$

If it takes a boat twice as long to go up and downstream as to go across a river current, traversing equal distances, what would be the ratio of the current speed to the boat's water speed?

Answer: $\sqrt{3}/2$

3–5. Michelson and Morley solved the problem of not knowing which way the ether current was flowing by rotating their apparatus. Explain.

SUGGESTIONS FOR FURTHER READING

COOPER, L. N. *An Introduction to the Meaning and Structure of Physics,* short edition. New York: Harper & Row, Publishers, 1970. (Chapter 23.)

MARCH, R. H. *Physics for Poets.* New York: McGraw-Hill Book Company, 1970. (Chapter 9.)

SMITH, J. H. *Introduction to Special Relativity.* New York: W. A. Benjamin, Inc., 1965. (Chapter 2.)

4

The Postulates of the Special Theory of Relativity

The result of the Michelson–Morley experiment was beyond doubt. Nature had revealed herself, or perhaps slammed the door in Michelson's face. Either way, the message was clear. The earth's motion relative to the ether was zero.

Attempts to Keep the Ether

To say that the ether does not exist would, perhaps, have been the most straightforward solution to the dilemma of the null observation. The physics community, however, could not readily dispense with an attitude that was so deeply ingrained. The reality of the ether had been too much a part of their universe.

It was possible that the Michelson–Morley experiment was not capable of detecting the ether for some unknown reason. But nothing was to be gained by that approach, for there was no physical basis for such a claim. The observation simply had to be taken at face value, and its implications understood.

Following the publication of the results of the Michelson–Morley experiment, there were several attempts to preserve the concept of a preferred ether frame. A first attempt was a length-contraction hypothesis, proposed independently in 1892 by two theoretical physicists, Lorentz and FitzGerald. Their ad hoc proposal was the idea that all bodies are contracted when in motion relative to the stationary ether, with the contraction taking place only in the direction of the relative motion. The light beam moving parallel to the ether current, therefore, would have a shorter distance to traverse, thus catching up with the beam going across the current. Such a contraction of length, however, was in no way to be regarded as a consequence of ether resistance. It was rather to be viewed as a "gift from above."

Another theoretical attempt was the ether-drag hypothesis. This hypothesis assumed that ether was dragged along by all objects so that they had no relative velocity with respect to the ether frame. It was only light that did not drag the ether with it, and light thus had the unique property of motion relative to the ether. This hypothesis, however, was readily amenable to experimental verification, and was rejected when it disagreed with observation.

Einstein's Principle of Relativity

Brushing aside all the ad hoc theories formulated to preserve the preferred ether frame of reference, Einstein suggested that Maxwell's electromagnetic theory be freed of the cumbersome ether hypothesis and that the notion of absolute motion be rejected as untenable in electromagnetism as in Newtonian mechanics.

The demise of absolute motion in electromagnetism was achieved by the introduction of a new principle of relativity. This new principle was more general than the Galilean version, so as to incorporate all physical phenomena. Thus, no physical experiment, of whatever nature, would establish absolute rest or absolute uniform motion, and the Michelson–Morley experiment was no exception.

In a more technical language, Einstein's principle of relativity may be stated as follows: *All the laws of physics must be so formulated as to be invariant (or the same in mathematical form) to all inertial frames of reference.* The new relativity principle is, indeed, an all-pervasive one and reflects Einstein's vision of a whole universe. Its implications are far-reaching as well as revolutionary. It serves as nothing less that the very foundation upon which a comprehensible universe would rest.

The Invariance of the Speed of Light

Maxwell's theory predicted a definite value of 3×10^8 meters per second for the speed of light in empty space. The empty space, however, was immediately clothed with a hypothetical ether so that the predicted speed of light could be interpreted as being relative to this ether frame of reference.

The experiment of Michelson and Morley entailed measuring the speed of the laboratory frame of reference relative to the ether frame. Their observation yielded the astonishing result that the relative velocity between the ether and any physical frame of reference is zero. It then follows that all observers, regardless of their relative velocities amongst

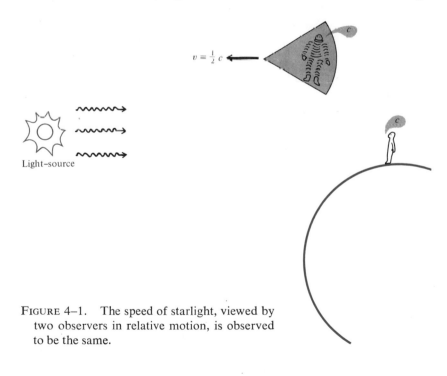

FIGURE 4–1. The speed of starlight, viewed by two observers in relative motion, is observed to be the same.

The Postulates of the Special Theory of Relativity _____

themselves, would measure the speed of light in vacuum to be the same as the value predicted by Maxwell's theory. The speed of light then becomes a universal quantity, and the concept of the ether frame fades as superfluous.

It is thus an experimental fact that the speed of light is the same to all observers, regardless of their relative state of motion. To impress on the reader the revolutionary character of this fact, let us consider an example. Suppose starlight is observed by an earthbound observer and an astronaut in a spaceship moving toward the star at half the speed of light. If they both measured the speed of the starlight, they would find the value to be precisely the same! An ordinary object, on the other hand, has different speeds relative to different observers. A stewardess, for example, who walks at a speed of 3 m.p.h. according to a passenger, walks at a fantastic speed of 603 m.p.h. to a ground-based observer watching the jet. Her walking speed is different to different observers, but not so with light. Light has the same velocity to all observers!

The speed of light is also independent of the velocity of its source. A laser beam fired from a fast-moving spaceship has the same velocity as one fired from a ground-based laser.

So one does not ever need to specify a frame of reference for the speed of light. It is the same to all observers, and all observers are equivalent insofar as the speed of light is concerned. This experimental fact is referred to as the principle of the invariance of the speed of light.

The Special Theory of Relativity

In 1905 Albert Einstein (1879–1955) startled the world with a new vision of the universe. He proposed that the universe be viewed in the light of two new principles:

1. All inertial frames of reference are equivalent.
2. The speed of light is the same to all inertial observers.

He considered these principles to be of such fundamental significance that he elevated them to the status of postulates for a physical theory. What emerged from these two postulates is the special theory of relativity.

The special theory of relativity provides a new framework upon which the physical models of the universe are to be constructed. It challenges

Physics: The Fabric of Reality⎯⎯⎯⎯⎯⎯⎯⎯⎯⎯⎯⎯⎯⎯⎯⎯⎯⎯⎯⎯

the absolute character of space and time, which was the foundation of Newtonian physics, and, in so doing, renders Newtonian physics inadequate to yield a correct picture of the universe. The new picture of the universe is thus to be drawn on the new canvas of relativistic space–time.

QUESTIONS

4–1. Justify the claim that ether does not exist in terms of the Michelson–Morley experiment.

4–2. If the existence of the ether had been established in another experiment, how might one interpret the result of the Michelson–Morley experiment?

4–3. Deduce the nonexistence of the ether from Einstein's principle of relativity?

4–4. Einstein said that the two postulates of the special theory of relativity are "only apparently irreconcilable," Deduce, then, the constancy of the speed of light from the principle of relativity.

4–5. What is the difference between Galileo's and Einstein's version of the principle of relativity?

SUGGESTIONS FOR FURTHER READING

COOPER, L. N. *An Introduction to the Meaning and Structure of Physics*, short edition. New York: Harper & Row, Publishers, 1970. (Chapter 24.)

VALLENTIN, A. *The Drama of Albert Einstein*. Garden City, N.Y.: Doubleday & Company, Inc., 1954.

The Postulates of the Special Theory of Relativity———————————

5

The Concept of Relative Time

Physical phenomena are the pictures that nature paints on the canvas of space and time. The special theory of relativity deals most decisively with the nature of this canvas. In this chapter, then, we shall discuss a component of this canvas, namely, time.

Concept of Absolute Time

"Absolutely, true, and mathematical time, of itself, and from its own nature, flows equably without any relation to anything external. . . ." So wrote Sir Isaac Newton. And so has man believed—that time has existed from the beginning, and that it hastens on whether man wills it or not. Man has, thus, tacitly assumed that there exists a cosmic clock, perhaps deep in outer space, ticking away the moments of true, absolute time.

Any notion of a cosmic clock is founded on the concept of absolute time. Deviations of the local time from absolute time are then manifestations of imperfect clocks. As clocks are perfected, local time approaches harmony with universal time. In the words of Newton himself, "relative time always more nearly approaches absolute time as we refine our measurements. . . ."

The absolute character of time is thus deeply ingrained in our way of thinking. We take it for granted that all reliable clocks run at the same rate, as they, in turn, tick in perfect harmony with the cosmic clock. Who of us, for example, would hesitate to set our watches on a jet flight when the pilot announces the time in the place of destination? For do we not always take it for granted that clocks, regardless of their respective states of motion, run at the same rate? May we not claim that such an assumption is justified, because we have yet to encounter a contrary experience?

Concept of Relative Time

Indeed, ordinary experience agrees with the concept of absolute time. But could it be that nature might be more subtle than our ordinary experience would reveal?

Long before there was any experimental evidence against the notion of absolute time, Einstein pondered deeply on the nature of time. He discovered that the concept of absolute time was not consistent with the postulates of relativity. He, however, considered the postulates of relativity as fundamental principles governing the entire universe, and thus judged the notion of absolute time to be untenable. Such was his faith in the power of pure thought.

The aspect of time at issue here is the notion of a time interval, because what is observable about time is the intervals of time. Time intervals are measured by a clock. A clock is anything which executes a rhythmic motion or has regular beats. A pendulum and a heart are familiar clocks.

The type of clock best suited to the discussion of the nature of time is a light-beam clock. A light-beam clock is a device that traps a light pulse between two parallel mirrors so that the pulse bounces off the mirrors at

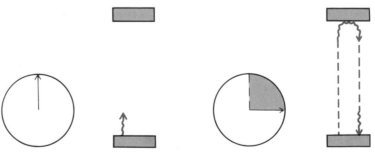

FIGURE 5–1.   An ordinary clock side by side with a light-beam clock.

Physics: The Fabric of Reality⎯⎯⎯⎯⎯⎯⎯⎯⎯⎯⎯⎯⎯⎯⎯⎯⎯⎯⎯⎯⎯

FIGURE 5–2.   Two clocks: one on ground and the other in spaceship.

perfectly regular intervals of time. Such a clock may seem a bit arbitrary and peculiar, but as a device to measure intervals of time, it is completely equivalent to clocks of all types.

As a way of deducing the nature of time from the postulates of relativity, let us start with two similar light-beam clocks. One is placed on the ground and the other in a spaceship. As long as the spaceship remains stationary on the take-off strip, both light-beam clocks tick at exactly the same rate.

Let us now observe two clocks in relative motion by setting the spaceship in motion. The ground-based observer would then observe that the light beam in the spaceship clock follows a diagonal path as it bounces between the mirrors. A light beam, however, has the same speed no matter where it is. A longer path for light, then, implies a longer time of transit. Consequently, the ground-based observer would observe that his own light beam clock completes a cycle before the spaceship clock. The spaceship clock, thus, runs at a slower rate than his own clock.

This observation about the difference in clock rates is actually an observation about the fundamental nature of time itself, for clocks measure time. Time itself, therefore, must be viewed by the ground-based observer as running slower in a moving frame of reference. Put another way, time in the spaceship appears dilated to the ground-based observer. The pilot's heart, for example, beats slower than that of the ground-based observer.

The Concept of Relative Time——————————————————————

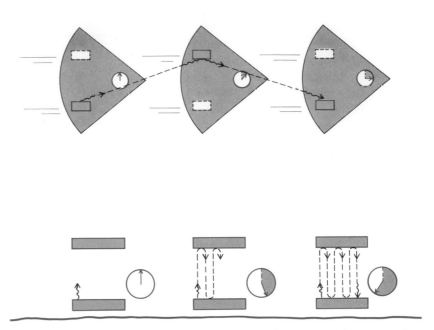

FIGURE 5–3.   The clock in a moving spaceship advances at a slower rate than the ground-based clock.

In fact, all activities in the spaceship proceed at a slower rate. Such dilation of time, however, arises strictly and entirely from the motion of the spaceship relative to the ground.

Time dilation is completely symmetrical among inertial observers. The spaceship observer would also observe time dilation in the ground-based frame of reference, as he would observe the ground to be in motion relative to the spaceship. In the theory of relativity, then, time exists in intimate relationship to the observer, and thus becomes an entirely relative concept.

Time Dilation—A Quantitative Analysis

Having thus gained a qualitative insight into the relative character of time, let us now seek to arrive at a quantitative relationship between two clocks in relative motion. Analysis is most straightforward if a light-beam clock is placed in the spaceship, and the light-beam clock is observed by both the spaceship and the earth-bound observers.

Suppose the spaceship observer finds that his clock advances by a time interval t' during a complete round trip of a pulse in the light-beam clock. If the length of the light-beam clock is L', the light pulse would traverse a distance of $2L'$ during the time interval t'. Or, in symbolic terms,

$$t' = \frac{2L'}{c}.$$

where c is the speed of the light pulse.

The earth-bound observer, on the other hand, must use his own clock as he observes the same bouncing light pulse in the light-beam clock aboard the spaceship. If his clock advances by a time t during a round trip of the pulse, the pulse must traverse a distance D, given by ct; that is, $D = ct$. The pulse, however, traces out a diagonal path, because by the time the pulse, starting at the bottom mirror, reaches the top mirror, the spaceship carrying the mirrors moves through a distance of $vt/2$. (v denotes the velocity of the spaceship relative to the earth.) When the pulse returns to the original bottom mirror, the spaceship moves through another distance of $vt/2$. Now if the vertical length of the light-beam clock is measured to be L, the distance traversed by the light pulse can be expressed in terms of L and v by use of the Pythagorean theorem. Con-

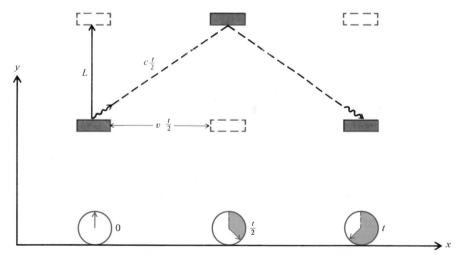

FIGURE 5–4. The light-beam clock aboard the spaceship, as observed by ground-based observer.

The Concept of Relative Time

sidering half of the total distance traversed, we have the following relationship:

$$\left(\frac{D}{2}\right)^2 = L^2 + \left(\frac{vt}{2}\right)^2$$

or

$$\left(\frac{ct}{2}\right)^2 = L^2 + \left(\frac{vt}{2}\right)^2 \qquad (1)$$

The question now arises: How is the time advance t registered on the earth-bound clock related to the time advance t' registered on the spaceship clock? Crucial to this question is the observation that the length of the light-beam clock is the same to both observers. (This will be proven in the following chapter.) Symbolically,

$$L = L' = \frac{ct'}{2}$$

Equation (1) can then be rewritten as

$$\left(\frac{ct}{2}\right)^2 = \left(\frac{ct'}{2}\right)^2 + \left(\frac{vt}{2}\right)^2$$

This expression, finally, reduces to

$$t = \frac{t'}{\sqrt{1 - \frac{v^2}{c^2}}}$$

which is the famous time-dilation equation derived by Einstein in his special theory of relativity.

Time Dilation—an Example

The concept of relative time seems to defy ordinary experience. To illustrate how bizarre it is, let us construct a hypothetical situation that would be consistent with the theory of relativity. Suppose the spaceship observer is an amateur horticulturist. He observes the life-span of a flower—a time interval from the moment a bud sprouts to the falling of the withered flower. He finds that his clock records the elapse of one month for the flower's life-span.

The earth-bound observer is also interested in knowing the life-span of the flower blooming in the spaceship. He first measures the speed of the spaceship to be 99.2 per cent the speed of light. According to the theory of relativity, then, he would observe the life-span of the flower to be 7.7 months, as deduced from the time-dilation equation:

$$t = \frac{1 \text{ month}}{\sqrt{1 - (0.992)^2}} = 7.7 \text{ months}$$

The earth-bound observer would, thus, enjoy the spaceship flower in bloom much longer than his spaceship counterpart.

Experimental Confirmation of Time Dilation

Early in the 1960's two American physicists, D. H. Frisch and J. H. Smith, embarked on the study of clocks in high-speed motion. (Their work was a pedagogical version of an earlier experiment performed by B. Rossi and D. B. Hall in 1941.) For their clocks they chose certain particles called muons. Muons are radioactive particles created in the atmosphere upon bombardment by cosmic rays. After a transient existence, they decay into other nonradioactive particles.

A radioactive substance can serve as a clock. This is so, because the amount of a radioactive substance changes with time according to a precise law. By measuring the radioactivity of a substance then, one can determine how much time has elapsed.

The radioactivity of a substance is specified by a half-life. A half-life is is defined as the time required for the decay of half of the original substance. If one begins with 1000 muons at a particular time, for example, there would be 500 muons still left after the elapse of one half-life. After the elapse of another half-life, there would be only 250 muons still left; the elapse of another half-life would leave 125 muons; and so on. The muon's half-life is 1.53 microseconds, or 1.53×10^{-6} seconds, when the muons are at rest.

Frisch and Smith wanted to measure the half-life of muons in flight. Of the muons in various states of motion, they chose to observe only those heading straight for sea level at a velocity of 99 per cent the speed of light. Their experiment consisted in counting the number of these particular muons at a certain altitude, and counting the surviving ones at sea level. Such a measurement would, upon analysis, yield the half-life of the muons traveling at 99 per cent the speed of light.

The Concept of Relative Time_____

They installed a muon detector at an altitude of 6000 feet on the summit of Mt. Washington in New Hampshire. The detector, adjusted to register only those particular muons, recorded the arrival of 563 such muons every hour.

Analysis on the basis of absolute time indicated that, at sea level, only

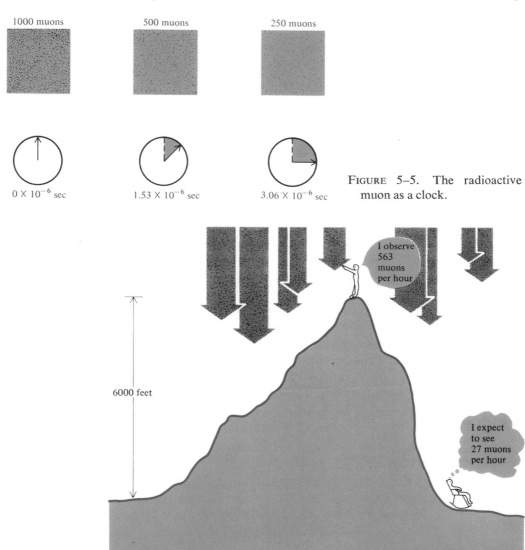

1000 muons   500 muons   250 muons

0×10^{-6} sec 1.53×10^{-6} sec 3.06×10^{-6} sec

FIGURE 5–5.   The radioactive muon as a clock.

I observe 563 muons per hour

6000 feet

I expect to see 27 muons per hour

FIGURE 5–6.   If time were absolute, 27 muons per hour would be counted at sea level.

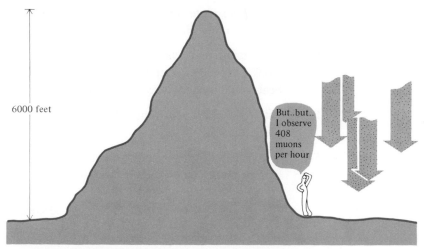

FIGURE 5–7.   Actual muon count at sea level: 408 muons per hour.

twenty-seven surviving muons would trigger the detector to be recorded every hour. On the other hand, the theory of relativity predicted that some 400 muons would trigger the detector at sea level every hour, as muons in flight would "live" longer. The experiment—in particular, its parent version—thus provided a decisive test for deciding whether time is absolute as assumed by Newton and by most of us, or whether time is relative as claimed by Einstein.

The apparatus was then transferred to Cambridge, Massachussetts, in order to count the muons at sea level. They again observed only those muons moving at 99 per cent the speed of light. The detector was turned on, and the muons started triggering the detector. After the elapse of one hour, there came the awaited answer. Four hundred eight muons had registered at the detector. Thus, the evidence was decisive that time is, indeed, a relative concept. Moving clocks do run slower than stationary clocks. Time exists not apart from but only in intimate relation to the observer.

QUESTIONS

5–1.   Give examples of your tacit assumption that time is absolute.

5–2.   The principle of the constancy of the speed of light and the con-

The Concept of Relative Time

cept of absolute time are mutually inconsistent. Why is Einstein justified in favoring the former and rejecting the latter?

5–3. On what grounds can one claim that because a moving light-beam clock advances slower that a stationary clock, the human heart-beat would also be subject to time dilation?

5–4. Prove the claim that time dilation is symmetrical among inertial observers.

5–5. If a spaceship travels in a straight line at 90 per cent the speed of light, what distance would it traverse during one hour as registered on the ground-based clock?

Answer: 9.7×10^{11} meters

5–6. If a clock is moving at a speed so that the factor $\sqrt{1-v^2/c^2}$ has a numerical value of 0.5, how much slower would it be advancing than a stationary clock?

Answer: slower by a factor of 2

5–7. If a spaceship is moving at a speed so that $\sqrt{1-v^2/c^2} = 0.9$, how much would the pilot age during the time that the ground-based observer ages one year?

Answer: 10.8 months

5–8. How would a relativist qualify the poetic image that "time flows like a river"?

SUGGESTIONS FOR FURTHER READING

COOPER, L. N. *An Introduction to the Meaning and Structure of Physics,* short edition. New York: Harper & Row, Publishers, 1970. (Chapter 24.)

MERMIN, N. D. *Space and Time in Special Relativity.* New York: McGraw-Hill Book Company, 1968. (Chapter 5.)

SMITH, J. H.: *Introduction to Special Relativity.* New York: W. A. Benjamin, Inc., 1965. (Chapter 3.)

6

The Concept of Relative Space

In Chapter 5 we analyzed the Frisch–Smith muon experiment from the viewpoint of a ground-based observer, concluding that the observed result could be interpreted as a confirmation of the time-dilation effect. But the question as to whether this particular interpretation could be taken as universally valid is a legitimate one, and requires a thorough analysis of the experiment in another inertial frame of reference.

The Frisch–Smith Experiment and the Muon Frame of Reference

It turns out that another convenient reference frame for an analysis of the muon experiment is the one attached to a spaceship moving in the same direction and at the same speed as the muons themselves. To the spaceship observer, the situation of the experiment appears as follows. The muons are now at rest, and the rest of the world is in motion. In particular, Mt. Washington and the Atlantic Ocean rush toward the spaceship at 99 per cent the speed of light. The experiment itself, however, remains the same; it still consists in counting the number of muons, initially at the moment when the mountain's summit brushes by the

39

spaceship, and later at the moment of "splashdown." Such a measurement would again yield an elapse of time between the summit and sea level.

The results to be obtained by the spaceship observer may be deduced from the principle of relativity. Since both are inertial observers, the ground-based observer and the spaceship observer would both agree on the number of muon counts at the summit and at sea level. The muon count would, therefore, indicate an elapse of 0.7 microsecond in the spaceship clock. (See Chapter 5.)

If the spaceship clock registers an elapse of only 0.7 microsecond, how far would the ocean rise toward the spaceship during that time? Since the distance traversed by a moving object is given by the product of the speed and the elapsed time, the answer is simple to calculate. It is 640 feet. In other words, the spaceship observer would measure the altitude of Mt. Washington to be only about 640 feet.

But did we not say that Mt. Washington is 6000 feet high? This question brings up an interesting point about the concept of absolute space. By absolute space we mean that a measurement of spatial separation would be the same to all observers, regardless of the state of the observer's motion. Applying it to the present situation, absolute space would imply that the ground-based observer and the spaceship observer

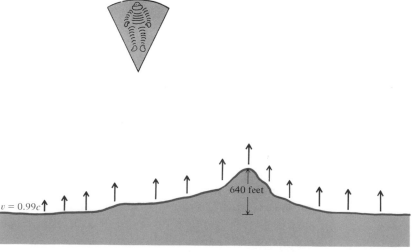

FIGURE 6–1.   Mt. Washington as viewed from the spaceship frame of reference.

Physics: The Fabric of Reality

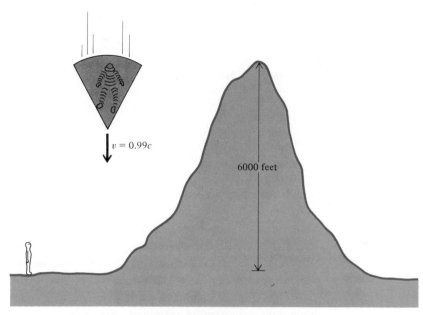

$v = 0.99c$

6000 feet

FIGURE 6–2.   Mt. Washington and the spaceship as observed by the ground-based observer.

would both agree on the altitude of the mountain. But, in reality, they do not agree!

The spaceship observer sees a dwarf of a mountain, only some 640 feet high. The idea of an absolute space, however familiar it may be, therefore, clearly contradicts observation. Thus emerges the notion of relative space; that the mountain, because it is in motion, appears contracted in altitude to the spaceship observer. This observation, in fact, may be generalized to include the contraction of all moving objects.

The interpretation of the Frisch–Smith muon experiment is subtle, indeed. One observer interprets it as a time-dilation effect, whereas another observer interprets it, with equal legitimacy, as a space-contraction effect. To all the other observers, still different from the ground-based and spaceship observers, the observed phenomenon would be interpreted as a mixture of time-dilation and space-contraction, suggesting that, in the theory of relativity, space and time do not play distinctly separate roles as they do in Newtonian physics. Rather, they merge to form a flexible stage to meet the increasing demands of nature's boundless imagination.

The Concept of Relative Space

Concept of Relative Space

The aspect of space, with which the theory of relativity deals, is the notion of a space interval, because what is observed about space is the intervals of space. Space intervals are measured with a meterstick.

Space has three dimensions. In a Cartesian coordinate system, the three dimensions of space are represented by three mutually perpendicular x, y, and z axes. The measurement of the size of a three-dimensional object would, then, require three metersticks which would be placed along the three axes.

Analysis of the relative nature of space is facilitated if the three dimensions of space are resolved into two dimensions relative to the direction of motion: the dimension parallel to the direction of motion referred to as the longitudinal dimension, and the two dimensions perpendicular to the direction of motion referred to as the transverse dimensions. The relative nature of the longitudinal and the transverse dimensions may then be examined separately.

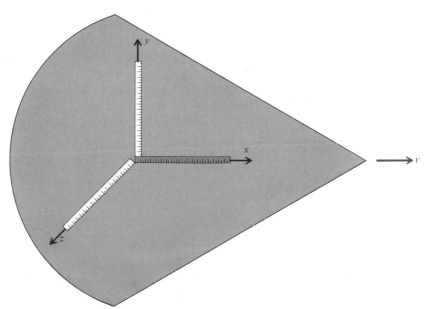

FIGURE 6–3.   A meterstick lying parallel to the direction of motion measures the "longitudinal" dimension of an object. Metersticks lying along the y and z axes measure the "transverse" dimensions of an object.

Physics: The Fabric of Reality

The Relative Nature of the Transverse Dimensions

Consider two similar metersticks, belonging to two observers Joe and Moe. Set the metersticks to stand parallel to each other along the y axis. As a way of comparing their lengths, place a paint brush at the top of Joe's meterstick, so that it would leave a paint mark on any object that is at least 1 meter long. A paint mark on Moe's meterstick, or its absence, would then indicate its length relative to Joe's meterstick.

At the outset, let us assume that the transverse dimensions of a moving object contract. Let us then ask if the conclusions deduced from such an assumption are consistent with the principle of relativity.

The assumption of transverse contraction leads Joe to make the following deductions: Moe's meterstick is moving, and therefore contracts. It thus misses Joe's paintbrush, receiving no stripe. According to Moe, on the other hand, it is Joe's meterstick that is in motion, and Joe's meterstick must, therefore, be shorter than his own. The result is that the paintbrush leaves a stripe on Moe's meterstick.

The results deduced from a common assumption, however, cannot be

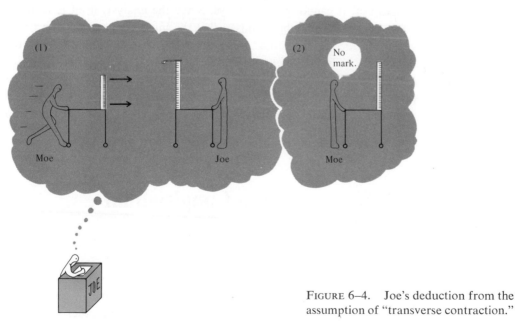

(1) Moe   Joe

(2) No mark. Moe

FIGURE 6–4. Joe's deduction from the assumption of "transverse contraction."

The Concept of Relative Space

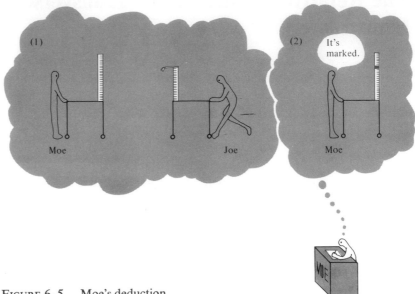

FIGURE 6–5. Moe's deduction.

contradictory. Moe's meterstick either ends up with a paint mark on it or it does not. It cannot be both ways. Since the analysis itself is logically consistent, it can only be surmised that the initial assumption of a transverse contraction must have been false. A correct assumption would then read: *No spatial contraction takes place in directions transverse (or perpendicular) to the direction of motion.* This particular method of proof is known as proof by contradiction. If we had started with an assumption of transverse expansion, we would again have been reduced to a contradiction.

The Relative Nature of the Longitudinal Dimension

In order to deduce the relative character of the spatial dimension that is parallel to the direction of motion, let us start with a set of two similar metersticks placed at right angles to each other. If one meterstick lies along the direction of motion, say the x axis, then the other meterstick

Physics: The Fabric of Reality

stands perpendicular to the direction of motion, or along the y axis. The question of interest here is how the lengths of the metersticks would compare if they are placed in a moving spaceship and observed from an earth-bound frame of reference. We already know that the "transverse" meterstick would remain the same whether in motion or not. Therefore, if the "longitudinal" meterstick is observed to be shorter than the "transverse" meterstick, the earth-bound observer would conclude that a "longitudinal" meterstick in motion is shorter than a similar stationary meterstick. He would further conclude that space contraction takes place only along the direction of motion.

The type of meterstick best suited to an analysis of this thought-experiment is a light-beam clock whose length is 1 meter. Two similar metersticks are then defined as two light-beam clocks which run at exactly the same rate. Two similar light-beam clocks are then set up in the spaceship as follows: One mirror of each clock is placed at the origin of a coordinate system. One light-beam clock is laid along the direction of motion of the spaceship, and the other is placed perpendicular to it.

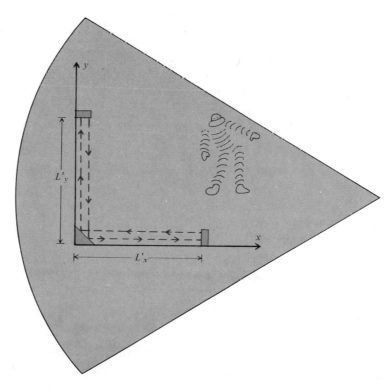

FIGURE 6–6. Two light-beam clocks set up in the spaceship, one along the x axis and the other along the y axis.

The Concept of Relative Space

Their equal lengths are then confirmed by observing that the light pulses in each cylinder leave and return to the mirrors at the origin at the same instants of time. If their lengths are denoted L'_x and L'_y, respectively, then

$$L'_x = L'_y$$

The length of the "transverse" clock, however, does not change whether the spaceship is in motion or not, and so can be denoted by another symbol, L_0, where the subscript 0 is chosen as a way of emphasizing that fact. The equal length of the clocks, according to the spaceship observer, may then be rewritten as

$$L'_x = L_0$$

Now let us set the spaceship in motion along the x axis with respect to the ground-based observer. The first question to arise is: Will the ground-based observer also find that the two light-beam clocks in the spaceship tick at the same rate? The answer must be in the affirmative, for otherwise the principle of relativity would be violated, making one observer distinguishable from the other.

But the ground-based observer would detect a time-dilation effect in the spaceship clocks. The spaceship clocks would run slower than the ground-based clocks. If the ground-based and spaceship observers register a time elapse of t_y and t'_y on their respective timing devices during one complete round trip of the pulse in the "transverse" light-beam clock, the relationship between them is given by the time-dilation equation:

$$t_y = \frac{t'_y}{\sqrt{1 - \dfrac{v^2}{c^2}}}$$

In terms of the length of the light-beam clock L_0, which is related to t'_y as $2L_0 = ct'_y$, the preceding expression for t_y may be written as:

$$t_y = \frac{2L_0}{c\sqrt{1 - \dfrac{v^2}{c^2}}}$$

Let us now denote by t_x the time elapse registered on the ground-based timing device during one complete round trip of the pulse in the light-

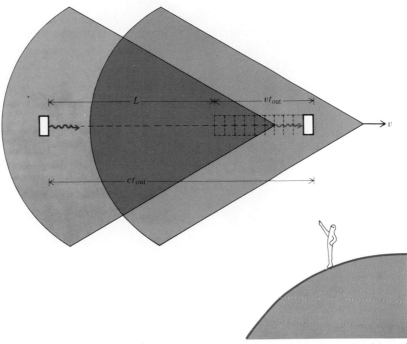

FIGURE 6–7. Light pulse in the spaceship, as observed by the ground-based observer: light pulse in transit from left to right.

beam clock lying along the x axis. We can then proceed to find an expression for t_x by analyzing the time-dilation effect of the "longitudinal" clock. During the transit of the light pulse from the left-hand mirror to the right-hand mirror, the pulse spends a time t_{out}, traversing a distance, ct_{out}. But the mirrors traverse a distance, vt_{out}, during the outbound flight of the pulse. The total distance that the pulse traverses is then the sum of L and vt_{out}, or symbolically,

$$ct_{out} = L + vt_{out}$$

where L is the length of the light-beam clock, as measured by the ground-based observer.

During the return trip the pulse spends a time, t_{in}, in transit, traversing a total distance of ct_{in}. But during the transit of the pulse, the left-hand mirror moves through a distance of vt_{in} toward the incoming pulse. The pulse, therefore, does not get to traverse the entire length L of the light-

The Concept of Relative Space

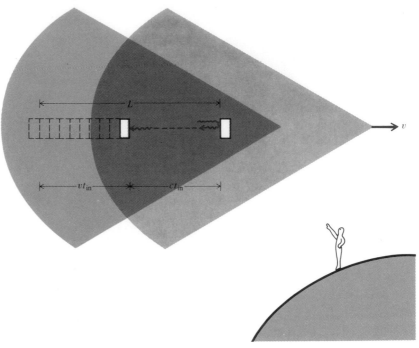

FIGURE 6–8. Light pulse in the spaceship, as observed by ground-based observer: light pulse in transit from right to left.

beam clock during the return trip. Rather, it traverses only a part of the length:

$$ct_{in} = L - vt_{in}$$

or

$$L = vt_{in} + ct_{in}$$

The total time that the pulse spends in making a round trip in the "longitudinal" clock is then the sum of t_{out} and t_{in}, or

$$t_x = t_{out} + t_{in}$$

In terms of the length, L, of the clock, a little algebra yields

$$t_x = \frac{2L}{c\left(1 - \dfrac{v^2}{c^2}\right)}$$

It has already been shown that the two clocks in our thought experiment must run at the same rate to all inertial observers as a consequence of the principle of relativity. This fact may now be summarized in the following simple equation:

$$t_y = t_x$$

or

$$\frac{2L_0}{c\sqrt{1 - \dfrac{v^2}{c^2}}} = \frac{2L}{c\left(1 - \dfrac{v^2}{c^2}\right)}$$

Canceling all common factors, this relationship reduces to

$$L = L_0\sqrt{1 - \frac{v^2}{c^2}}$$

where L and L_0 denote the lengths of the "longitudinal" clock in motion and at rest, respectively.

We have thus shown how length contraction follows directly from the postulates of the special theory of relativity. To state length contraction again: *Moving objects contract in the direction of their motion.* This is a truly profound result. Space, like time, must also be regarded as existing not apart from but only in intimate relation to the observer.

Length Contraction—An Example

The spatial dimensions observed by fast-moving observers are quite different from those we would ordinarily observe. Consider, for example, a frame of reference attached to an electron moving at 99.98 per cent the speed of light. Such high speeds can readily be attained, for example, in the 2-mile-long Stanford Linear Accelerator. The electron, however, does not have 2 miles to travel. To it the accelerator appears severely contracted, and according to the length contraction relation, the accelerator would appear to be only about 208 feet long:

$$L = 2 \text{ miles}\sqrt{1 - (0.9998)^2} = 0.04 \text{ miles}$$

or

$$L = 208 \text{ feet.}$$

The Concept of Relative Space

Length contraction plays an important part in the discussions of space travel. To a fast-moving spaceship the distance to a stellar object, for instance, would appear contracted, and the object, therefore, would not really be as far as an earth-bound observer would perceive it to be. Indeed, the relative character of space makes the universe a much more intimate place.

QUESTIONS

6–1. In the concepts of absolute space and absolute time, space and time are totally unrelated and independent of each other. Argue, using the example of the Frisch–Smith muon experiment, how space and time can no longer be regarded as unrelated, independent concepts.

6–2. Prove that the transverse dimensions of an object remain the same whether the object is in motion or at rest, by assuming that they expand when the object is in motion.

6–3. Suppose a light-beam clock, lying parallel to the x axis, is observed by an observer moving along the positive x direction.
    a. In going from one mirror to the other, in which direction would the light pulse spend more time?
    b. What would be the difference in the transit times of the light pulse?

Answer: $2vL/(c^2 - v^2)$

6–4. "What is the distance between Washington and Peking?" constitutes a meaningful question to one who accepts the validity of absolute space. Why is the question incomplete to one who rejects the notion of absolute space? How would the question be rephrased?

6–5. Suppose a star is 25 light-years away from earth, according to the earth-bound observer.
    a. Observed from a spaceship travelling from earth to the star at 99.98 per cent the speed of light, what would be the distance between earth and the star?
    (*Hint:* $\sqrt{1 - (0.9998)^2} = 0.02$)

Answer: ½ light-year

b. From another spaceship traveling at the same speed but in a different direction, namely, at 90 degrees, what distance would be observed between earth and the same star?

Answer: 25 light-years

6–6. If one observes a meterstick moving nearly at the speed of light, what would he observe about its length?

Answer: any value between 1 meter and nearly zero

6–7. If the distance between New York and San Francisco is 3000 miles according to ground-based observers, would the passengers on a jetliner in transit observe the distance to be appreciably shorter? Explain by doing rough calculations.

SUGGESTIONS FOR FURTHER READING

MARION, J. B. *A Universe of Physics*. New York: John Wiley & Sons., 1970. (Sections 4.2–4.3.)

MERMIN, N. D. *Space and Time in Special Relativity*. New York: McGraw-Hill Book Company, 1968. (Chapter 6.)

7

The Twin Paradox and Clock Synchronization

Space travel is no longer a purely academic matter. Man has been to the moon, and, at the time of this writing, Pioneer 11 is ready for launching on a twenty-two-month journey to probe Jupiter.

Perhaps the most awesome aspect of space venture is the fantastic distances involved and the correspondingly long time spans required. It is for these reasons that a spaceship for deep interstellar explorations must necessarily travel at nearly the speed of light. But at such speeds, relativistic effects dominate, and any discussion of space travel, apart from the theory of relativity, would be completely meaningless. By the same token, analysis of space travel provides an excellent opportunity to apply the relativistic concepts.

Formulation of the Twin Paradox

Joe and Moe are identical twins. On their twentieth birthday they set out on separate ventures. Moe is chosen to pilot the first spaceship on an interstellar voyage. Remaining on earth, Joe assumes a keen interest in his brother's journey into space.

The first question of mutual interest is: When would they be reunited on earth? Joe is provided with the information that Moe's destination is a star 25 light-years away, and that Moe's spaceship has been engineered to attain a speed 99.98 per cent of the speed of light relative to earth. (A light-year is the distance traversed by light in the time span of one year.)

Joe's analysis of the trip proceeds as follows: The star is 25 light-years away, and Moe is travelling at nearly the speed of light. During Moe's outbound journey, Joe himself ages twenty-five years, and another twenty-five years during the inbound trip. Moe, however, ages much less, because of the time-dilation effect. In fact, Moe ages only one year during the entire journey, as deduced below from the time-dilation equation:

$$T_{\mathrm{moe}} = T_{\mathrm{joe}} \sqrt{1 - \frac{v^2}{c^2}} = 50 \times 0.02 = 1 \text{ year}$$

So Joe expects to welcome his twin brother home when he is about seventy years of age. He further expects Moe to emerge from his spaceship as a twenty-one-year-old youth.

The trip is also analyzed by Moe. He argues, on the basis of time dilation and space contraction, as follows: As the star approaches him, the earth recedes, both nearly at the speed of light. The distance of separation L between earth and star contracts to 0.5 light-year according to the length-contraction equation:

$$L = L_0 \sqrt{1 - \frac{v^2}{c^2}} = 25 \times 0.02 = 0.5 \text{ light-year}$$

The star, therefore, reaches the spaceship in half a year, and the earth returns in another half year. Thus he concludes that he would age one

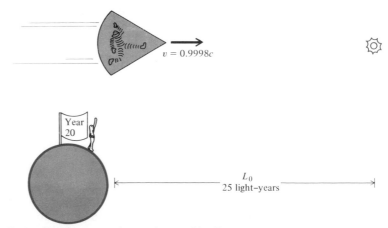

FIGURE 7–1.   Moe's space trip, as observed by Joe.

Physics: The Fabric of Reality

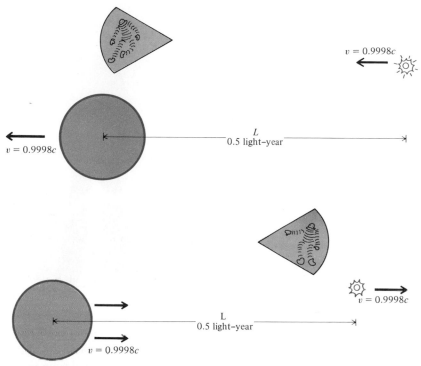

FIGURE 7 2.   As observed by Moe in the spaceship: (a) earth receding and
star approaching, (b) earth returning and star receding.

year during the entire trip. Both analyses, then, lead to the same conclusion as to the aging of Moe during the flight.

As Moe proceeds further in his analysis, he discovers that Joe would
age much less than he. Since it is Joe that is in motion, Joe's time must
dilate. He deduces that Joe would age only about a week during his year
away in space:

$$T_{\text{joe}} = T_{\text{moe}}\sqrt{1 - \frac{v^2}{c^2}} = 1 \text{ year} \times 0.02 \cong 6 \text{ days}$$

If he expects, on the basis of his present analysis, to meet a youthful
Joe upon returning to earth, Moe would be caught by complete surprise.
The brother whom he embraces at the spaceport would actually be a
white-haired seventy-year-old man.

Moe's present analysis of the trip is apparently not quite correct, as it
fails to prepare him for the reality of the spaceport encounter. There is

The Twin Paradox and Clock Synchronization⎯⎯⎯⎯⎯⎯⎯⎯⎯⎯⎯

FIGURE 7–3.   Spaceport encounter.

no question, however, that his analysis of the effects of time dilation and length contraction is flawless. A reasonable resolution of the dilemma is then to suggest that Moe's erroneous conclusion must have followed from an incomplete knowledge of relativity.

Re-examination of the Twin Paradox

The first area to examine is whether Moe is a legitimate inertial observer. It turns out that Moe is, indeed, distinguishable from Joe—not during the coasting part of the voyage, but during the brief periods of acceleration and deceleration. To attain a speed of 0.9998c, Moe must first accelerate. Likewise, he must decelerate in order to land. Effects of acceleration and deceleration, however, are readily detectable, contrary to the nondetectable effects of uniform motion.

A simple test of accelerated motion may be provided by a pendulum. A freely suspended pendulum would remain suspended if the spaceship is in a state of uniform motion. If the spaceship is accelerated forward, however, the pendulum would swing backward. One could thus determine that one is in a state of accelerated motion. Treatment of noninertial observers falls in the domain of Einstein's general theory of relativity. Einstein showed that an accelerated clock runs slower than a clock at rest.

It is, then, obvious that Moe would suffer an absolute dilation of time during the periods of acceleration and deceleration. But the slowing down of the spaceship clock because of acceleration is the same in magnitude whether Moe visits a nearby object or a distant one. His total aging during the flight, therefore, depends primarily on the distance of the trip, so that the effect of acceleration on Moe's time can be made arbitrarily small by increasing the length of the trip. Moe's neglect of the effect of acceleration in his analysis can, therefore, be justified. But this means that the source of the discrepancy that appears in Moe's analysis has yet to be identified and must be sought elsewhere.

The Relative Nature of Clock Synchronization

In his analysis of the trip, Moe tacitly assumed that the earth and star clocks, synchronized to Joe, remained synchronized, also, to him during the entire trip. But is an assumption such as absolute clock synchronization valid in relativity?

To begin, let us define what is meant by the statement that two clocks which are separated in space are synchronized. In order to synchronize two clocks one must send a message from one clock to the other, e.g., by means of a light signal. Since light travels at a finite speed, it is not possible to read off the time in two spatially separated clocks at the same instant of time. In ordinary experience this fact is of no consequence, as the spatial separations one encounters are not large and the speed of light is almost infinite by comparison. For example, it would take the light 10^{-8} seconds to go from one clock to another that are 3 meters apart. But a precise definition of clock synchronization is, nevertheless, essential in arriving at a correct picture of the universe.

Two clocks, which are separated by a distance of L_0 and at rest with respect to each other, may be synchronized by an observer stationary relative to the clocks in the following way. Let us say that Clock 1 is equipped with a light emitter and Clock 2 with a light receiver, and that they are 5 light-hours apart. Then let Clock 1 read 0 o'clock when a flash is emitted. Set Clock 2 such that it would read 5 o'clock when it receives the flash. The two clocks are then clearly synchronized, for one could claim that since the light flash spends 5 hours in transit from Clock 1 to Clock 2, Clock 2 must have read 0 o'clock at the moment that Clock 1 emitted the flash. Both clocks, of course, advance at the same rate.

Let us now look at the same two clocks from the viewpoint of a moving observer, Moe. Moe would observe that both Clock 1 and Clock 2 move

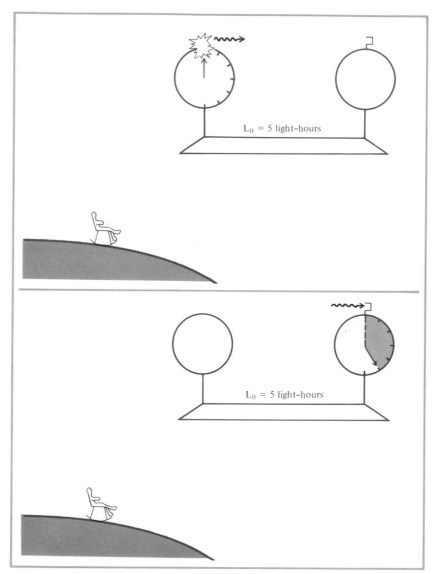

FIGURE 7–4. Two clocks synchronized according to Joe.

toward him at a velocity v. The distance of separation, L, between the clocks would then appear contracted according to

$$L = L_0 \sqrt{1 - \frac{v^2}{c^2}}$$

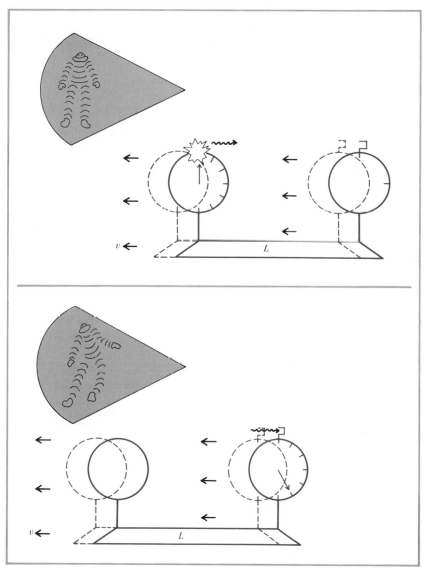

FIGURE 7–5. Two clocks synchronized relative to Joe—now observed by Moe.

The Twin Paradox and Clock Synchronization

As a light flash transits from Clock 1 to Clock 2, the clocks maintain their uniform motion toward Moe. During the transit of the flash Moe's clock in the spaceship would advance by a certain time interval, which we

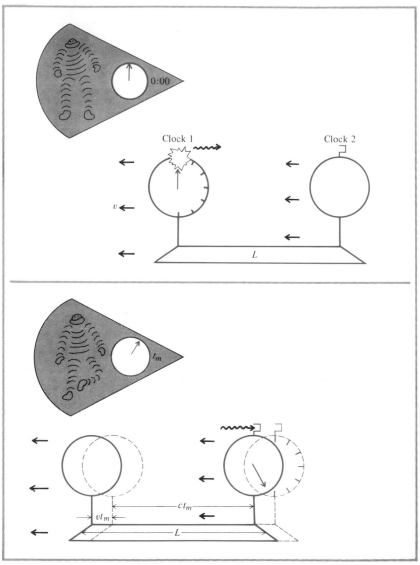

FIGURE 7–6.   During the transit of the light signal from Clock 1 to Clock 2, Moe's clock advances a time interval, t_m, so that the light traverses a distance of ct_m and the clocks move through a distance of vt_m.

Physics: The Fabric of Reality

shall denote by t_m. Then, from Figure 7-6, the following relationship results:

$$L = ct_m + vt_m$$

or

$$t_m = \frac{L}{c + v}$$

But Clock 1 and Clock 2 advance slower than Moe's own timing device fixed in the spaceship, because they are in motion relative to the spaceship. If we denote by t_{adv} the amount of time by which Clock 1 and Clock 2 actually advance between the two events of light-emission in Clock 1 and light-absorption in Clock 2, t_{adv} is, then, related to the time elapse on Moe's timing device via the time-dilation equation:

$$t_{adv} = t_m \sqrt{1 - \frac{v^2}{c^2}}$$

or, substituting for t_m,

$$t_{adv} = \frac{L \sqrt{1 - \frac{v^2}{c^2}}}{c + v}$$

This expression, after a few algebraic steps, reduces to

$$t_{adv} = \frac{L_0}{c} - \frac{L_0 v}{c^2}$$

Let us now seek to interpret the last expression. Recall that, according to the stationary observer, Joe, Clock 1 and Clock 2 both advance by a time-interval given by L_0/c between the events of light-emission by Clock 1 and light-absorption by Clock 2. According to Moe, on the other hand, the same two clocks advance less by an amount given by $L_0 v/c^2$ between the same two events.

In order that we may be specific, let us take the velocity of the spaceship to be $0.8c$ relative to the clocks. Then Moe would observe that both Clock 1 and Clock 2 advance by one hour during the transit of the light flash from Clock 1 to Clock 2, as easily computed from

$$t_{adv} = \frac{L_0}{c} - \frac{L_0 v}{c^2} = 5 \text{ hr} - 4 \text{ hr} = 1 \text{ hr}$$

The Twin Paradox and Clock Synchronization

Joe, on the other hand, would observe that the same two clocks advance by 5 hours during the transit of the light flash. That Joe and Moe measure different time intervals between light emission and light absorption comes as no surprise. But it does raise a question of great significance: How does

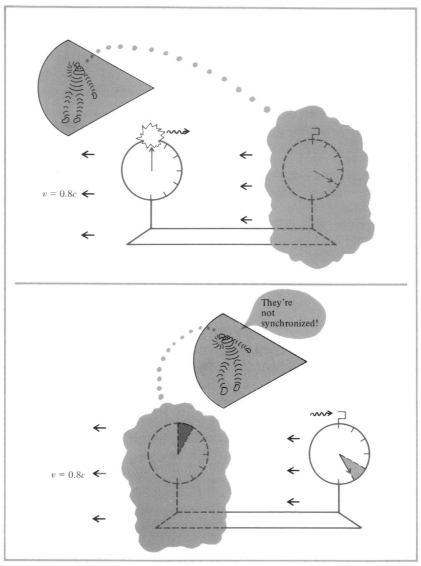

FIGURE 7–7.   Moe deduces that when Clock 1 reads 0 o'clock, Clock 2 reads 4 o'clock, and that when Clock 2 reads 5 o'clock, Clock 1 reads 1 o'clock.

Physics: The Fabric of Reality

the difference in their time measurements affect the state of synchronization between Clock 1 and Clock 2? When they start out synchronized to Joe, do Clock 1 and Clock 2 remain synchronized to Moe?

Both Joe and Moe agree that when it emits a flash, Clock 1 reads 0 o'clock. Similarly, they also agree that when it receives the flash, Clock 2 reads 5 o'clock. This mutual agreement follows directly from the principle of relativity, as may readily be demonstrated by the use of the following device. Let Clock 1 print out the numerical value of the hour when it emits a light pulse. This number would be 0. Let Clock 2, likewise, print out the numerical value of the hour when it receives a pulse, which would be 5. These print-outs could then be examined by any observer, with a resulting agreement among all observers.

Knowing that both Joe and Moe agree that Clock 2 reads 5 o'clock when it receives the flash, we may now ask what Moe would deduce about the hour on Clock 2 at the moment Clock 1 reads 0 o'clock. Recall that since Clock 2 advances 5 hours during the transit of the light flash, Joe would deduce Clock 2 to have read 0 o'clock at the moment of light emission by Clock 1. After all, these clocks were set so as to be synchronized to Joe. According to Moe, on the other hand, Clock 2 advances by only 1 hour. Moe would, therefore, deduce that Clock 2 must have read 4 o'clock at the moment Clock 1 emitted a light flash. But at the moment of light emission, Clock 1 read 0 o'clock. The two clocks are, then, clearly not synchronized. Rather, Clock 2 is *ahead* of Clock 1 by 4 hours, according to Moe.

In general, if a stationary observer finds two clocks in synchronization, a moving observer would find that: *The "rear" clock is ahead of the "front" clock by a time given by* L_0v/c^2. This is referred to as the relativity of clock synchronization.

Resolution of the Twin Paradox

Let us now apply the relativity of clock synchronization to the discussion of the twin paradox. According to Joe, the earth clock is synchronized with the star clock. When it is Year 20 in the "year of Joe" on earth, it is also Year 20 on the star. To Moe, however, the clocks are not synchronized, one being ahead of the other. Just which one is ahead depends on whether Moe is inbound or outbound. During the outbound flight, for instance, Moe would observe the star approaching him and the earth receding away from him. The star clock would then constitute a "rear" clock and thus would be ahead of the earth clock. Consequently, Moe's analysis of the trip must be revised so as to incorporate the concept of relative clock synchronization:

Moe takes off in the Year 20 on earth. The moment his spaceship attains the coasting speed, he asks to know what year it is on the star. On the star, it is Year 45, since the star clock, being the "rear" one, is ahead of the earth clock (which, in this case, is the "front" clock) by 25 years, which is obtained from

$$\frac{L_0 v}{c^2} \cong \frac{25 \text{ light-years}}{c} \cong 25 \text{ years}$$

When Moe stops briefly on the star, it is, of course, Year 45 on earth also, plus about three days, because about three days elapse both on earth and on the star during the outbound trip.

As he hops on the inbound spaceship moving at $v = 0.9998c$, Moe again asks to know what year it is on earth, and deduces that the earth clock (now a "rear" clock) must be ahead of the star clock (the "front" clock) by twenty-five years. Life on earth advances by about three days during the return trip, so when Moe finally lands at the spaceport, the earth

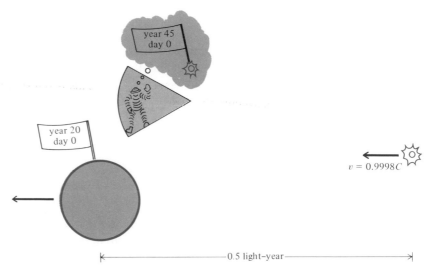

year 45
day 0

year 20
day 0

$v = 0.9998c$

0.5 light-year

FIGURE 7–8. Aboard the spaceship, Moe immediately asks to know what year it is on the star. It is Year 45, day 0.

clock would show Year 70 and Day 6. Joe must, therefore, be about seventy years old, and, of course, Moe himself is twenty-one years old, since the entire trip took about a year. The twin paradox is thus resolved.

It is, indeed, the relativity of clock synchronization that enables one to complete a long interstellar journey. To the earth-bound people, Moe's trip takes some fifty years, but to Moe it lasts only one year. He still has a lifetime ahead of him to begin an earth-bound career after he returns from his space ventures.

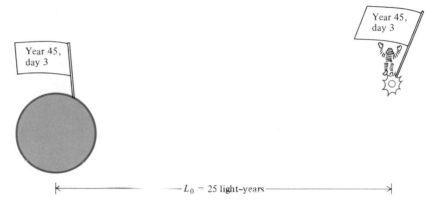

FIGURE 7–9.   Moe lands on the star in the Year 45, day 3.

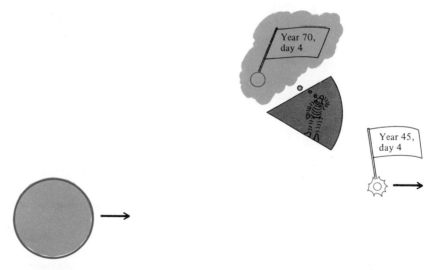

FIGURE 7–10.   Upon boarding the inbound spaceship, Moe finds that it is Year 45, day 4 on earth.

The Twin Paradox and Clock Synchronization

Simultaneity

A notion related to clock synchronization is that of simultaneity. Two events that are separated in space may be simultaneous to one observer but not simultaneous to another in relative uniform motion.

Suppose Moe sits at the midpoint of his spaceship, and the ends of the spaceship get hit by lightning at the moment he is directly above ground-based Joe. Joe would then observe that the lightning bolts hit the spaceship at the same time, thus constituting two simultaneous events. But Moe would observe that a signal from the front end of his spaceship reaches him before one from the rear. These two events, therefore, do not occur simultaneously to Moe. It is thus clear that no absolute significance can be attached to the concept of simultaneity.

QUESTIONS

7–1. How much time would a spaceship spend traversing a distance of 1 light-year, if it coasted at 10 per cent the speed of light?

Answer: Ten years

7–2. Take the distance between New York and Los Angeles to be about 3600 miles, and suppose that New York's standard clock is synchronized to Los Angeles' standard clock, according to the ground-based observer. To an observer aboard a jetliner cruising at 600 m.p.h. relative to the ground, which city clock would, in principle, be ahead, and by how much?

Answer: Los Angeles, by an undetectable amount

7–3. If Moe's spaceship were coasting at 80 per cent the speed of light, instead of $0.9998c$ as in the text, how much would Moe observe the earth and the star clocks to be out of synchronization?

Answer: Twenty years

7–4. If Moe, the hero of the space venture story in the text, had a one-year-old daughter when he departed, how old would she be when he returned to earth?

Answer: Fifty-one years old

7–5. Suppose a train is too long to be completely contained in a tunnel when it is at rest. Suppose, further, that at a certain relativistic speed, the train is observed to be just contained in the tunnel by a ground-based observer. How would an observer on the train describe its passage through the tunnel? Resolve the difference in the two observations.

7–6. In discussing the relative nature of simultaneity, we considered the example of lightning bolts striking the ends of a spaceship. We concluded that Moe would observe the front end of the spaceship to be hit before the rear end. Deduce Moe's observation from the relativity of clock synchronization.

SUGGESTIONS FOR FURTHER READING

MERMIN, N. D. *Space and Time in Special Relativity*. New York: McGraw-Hill Book Company, 1968. (Chapter 16.)

SMITH, J. H. *Introduction to Special Relativity*. New York: W. A. Benjamin, Inc., 1965. (Chapter 6.)

The Twin Paradox and Clock Synchronization_____

8

$$E = mc^2$$

Perhaps the most famous consequence of the principle of relativity is that mass and energy are equivalent concepts. It is this equivalence, for instance, that lies at the heart of nuclear energy. Far-reaching as it is, the mass-energy equivalence can, amazingly enough, be deduced from a simple relativistic thought experiment.

Observations on a Light Emitter

Consider a light emitter fixed in a spaceship piloted by observer Moe. Let the only stipulation be that the emitter send out two light pulses of equal energy in opposite directions. Now the energy of a light pulse is proportional to its frequency, which may, in the compact notation of mathematical symbols, be expressed as

$$E = hf$$

where E and f denote energy and frequency, respectively, and h represents a proportionality constant. The stipulation that both pulses carry the same energy then implies that Moe would measure the frequencies of the light pulses to be equal. Using a primed coordinate system, he would summarize his result in the following simple equation:

$$f_1' = f_2'$$

where the subscripts 1 and 2 refer to the forward and the backward pulses, respectively.

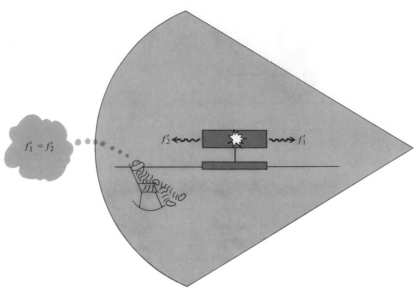

FIGURE 8–1. Two light pulses of equal frequency, moving in opposite directions.

In order to view the same light emitter from another frame of reference, let us set the spaceship moving at a constant velocity, v, relative to stationary observer Joe. Let us then ask how much energy the spaceship emitter would generate in the form of pulses, as observed by Joe. Joe's observation would consist in measuring the frequencies of the pulses. One immediate result would be that the two light pulses do not have the same frequency. This follows from a rather familiar property of waves called the Doppler effect.

The Doppler effect refers to the observation that the frequency of a wave that one observes depends upon the relative velocity between the source of the wave and the observer. The pitch of a horn sound from a passing car, for example, rises and falls, as the car approaches and then recedes from an observer. This change in frequency is due entirely to the relative velocity between the car and the observer. In general, a wave from a source approaching an observer rises in frequency, and a wave receding from an observer falls in frequency.

It is due to a Doppler effect that Joe would measure the frequency of the approaching pulse to be different from that of the receding pulse. Compared to Moe's measurements, Joe would find that the approaching

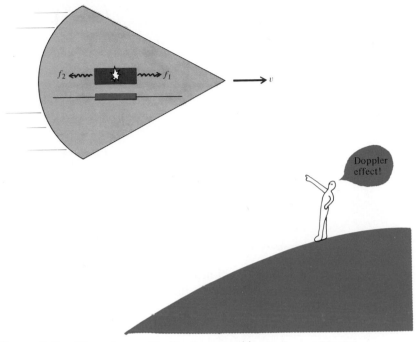

FIGURE 8–2. The pulses emitted in the spaceship, as observed by Joe.

pulse has a higher frequency, and the receding pulse a lower frequency, than Moe's values. In algebraic notations,

$$f_1 > f_1'$$

$$f_2 < f_2'$$

where the unprimed frequencies refer to Joe's measurements.

The important question now is: Would the total energy of the pulses be the same to both observers? A common-sense response would be to argue in the affirmative. For it would seem only too reasonable that a light emitter should generate the same amount of energy for all observers. But we have learned to be cautious with common sense in relativity. So let us proceed by computing the sum of the measured frequencies of the pulses, i.e., $(f_1 + f_2)$ and $(f_1' + f_2')$. Now these sums can be compared if we know how Joe's measurements are related to Moe's. The relationship is given by the so-called relativistic Doppler equations:

$E = mc^2$ _____

$$f_1 = f_1' \frac{\left(1 + \dfrac{v}{c}\right)}{\sqrt{1 - \dfrac{v^2}{c^2}}}$$

$$f_2 = f_2' \frac{\left(1 - \dfrac{v}{c}\right)}{\sqrt{1 - \dfrac{v^2}{c^2}}}$$

Adding both sides of the equations, we obtain

$$f_1 + f_2 = \frac{f_1' + f_2'}{\sqrt{1 - \dfrac{v^2}{c^2}}}$$

Thus the sums of measured frequencies are not equal.

In terms of energy, the preceding expression takes the form:

$$E = \frac{E_0}{\sqrt{1 - \dfrac{v^2}{c^2}}}$$

where E_0 and E denote the total energy of the light pulses, as observed by Moe and Joe, respectively. The subscript zero is to stress the fact that the emitter is at rest with respect to Moe. This equation says that the energy observed by Joe is greater than the energy observed by Moe; that is, $E > E_0$. But how could the same emitter produce more energy for one observer than for another? This is a disturbing result, and it clearly contradicts the familiar version of the principle of the conservation of energy: Energy can be neither created nor destroyed; energy can only be transformed from one form to another.

Einstein's Analysis

To resolve this peculiar observation, Einstein suggested that we inquire into the very source of the light pulses. The only aspect of the light emitter that is different to the two observers is its state of motion. The emitter is moving relative to Joe, whereas it is not moving relative to

Moe. So, Einstein argued, the "extra energy" observed by Joe must come from the energy of motion.

The energy of motion is called kinetic energy. Any object that is in motion has kinetic energy. If an object is moving slowly compared to the speed of light, the expression for kinetic energy takes the form:

$$\text{K.E.} = \tfrac{1}{2} m_0 v^2$$

where m_0 is the mass of the object at rest and v is its velocity.

By mass we mean the inertia of an object—its resistance to change in motion. The higher the mass of an object, the more energy it takes to change its motion, i.e., its velocity. Conversely, if there are two objects of different masses moving at the same velocity, the one of greater mass is able to deliver a greater energy.

Since kinetic energy depends on mass and velocity, the "extra energy" observed by Joe must also depend on those two quantities. But the "extra" portion of the energy can be made to disappear by stopping the light emitter. However, an emitter at rest can still emit energy, namely, E_0, which is just the energy observed by Moe. Therefore, the real source of energy must be the other variable quantity of the kinetic energy, namely, mass itself. But if mass is the source of energy, asked Einstein, what must be the relationship between mass and energy?

The Mass–Energy Relationship

Let Δm denote the portion of the mass of the emitter that is converted into the energy of the light pulses. Then the kinetic energy of this bit of mass is $\tfrac{1}{2} \Delta m v^2$, which must be equal to the "extra energy" observed by Joe.

In order to facilitate our analysis, let us make the approximation that the velocity of the light-emitter (and the spaceship) relative to Joe is much smaller than the speed of light. In that approximation, the relationship between E and E_0 takes on a simpler form:

$$E = E_0 \left[1 + \tfrac{1}{2} \left(\frac{v^2}{c^2} \right) \right]$$

so that the "extra energy" is

$$E - E_0 = \tfrac{1}{2} E_0 \left(\frac{v^2}{c^2} \right)$$

$E = mc^2$

But the "extra energy" equals the kinetic energy of the mass, Δm. That is,

$$\tfrac{1}{2}\, E_0\!\left(\frac{v^2}{c^2}\right) = \tfrac{1}{2}\,\Delta\, mv^2$$

or

$$E_0 = \Delta mc^2$$

This is the famous mass–energy equation of Einstein.

Let us now recapitulate Einstein's arguments. Einstein requires that the general idea of energy conservation be preserved in the theory of relativity. But in so doing, he finds it necessary to modify the more familiar version of energy conservation, i.e., energy can be neither created nor destroyed. The only way to interpret the "extra energy" of the light emitter within the principle of energy conservation is to view mass as the source of energy. In this perspective, then, energy can be created or destroyed, but only with a simultaneous destruction or creation of mass. Energy and mass thus become equivalent concepts. When mass is converted into energy, the energy released simply equals the mass multiplied by the square of the speed of light.

$$E = mc^2$$

The Relative Nature of Mass

Mass, now being equivalent to energy, must then be as much of a relative concept as energy. The equation

$$E = \frac{E_0}{\sqrt{1 - \left(\dfrac{v^2}{c^2}\right)}}$$

which describes the relative nature of energy, may equivalently describe the relative nature of mass:

$$m = \frac{m_0}{\sqrt{1 - \left(\dfrac{v^2}{c^2}\right)}}$$

where m denotes the mass of an object in motion and m_0 denotes the mass of the same object at rest. In words, the equation says that *moving mass increases*.

To illustrate the relativity of mass, let us compute the mass of an object when it is moving at 90 per cent the speed of light. If its mass at rest is 1 gram,

$$m = \frac{1 \text{ gram}}{\sqrt{1 - (0.9)^2}} = 2.2 \text{ grams}$$

A further calculation shows that to push an object from $v = 0.90c$ to $v = 0.99c$ it would take about four times as much energy as to push it from $v = 0$ to $v = 0.90c$. It is this fact that, perhaps, presents the greatest technological challenge in constructing a spaceship that would attain velocities approaching the speed of light. Even pushing tiny electrons to near the speed of light requires giant "atom smashers," whose construction costs are considerable.

Another implication of the relative character of mass is that the speed of light serves as the ultimate speed limit for all objects. The mass of an object becomes infinite at the speed of light, but infinite mass is both unattainable and unphysical. No object, therefore, can travel at the speed of light.

Mass as a Source of Energy

The equivalence of mass and energy is, indeed, an inescapable consequence of the principle of relativity. A gram of matter, for instance, would, in principle, convert into as much as 20 trillion calories of energy. The utilization of matter as a source of energy, however, was long regarded only as a theoretical possibility. It was some thirty years after Einstein's original work that scientists stumbled on the observation that uranium undergoes fission. Nuclear fission incurs a loss of mass—typically of the order of 0.1 per cent the mass of the splitting nucleus—but as small as the mass loss is, the energy converted therefrom is enormous. Nuclear fission is one instance where nature uses $E = mc^2$.

The mass of matter is concentrated in the nucleus. To harness the energy of mass, then, man must learn to harness the nucleus itself—a task that is proving to be formidable. Nevertheless, the ultimate solution to the growing energy needs of a technological world may still lie in the successful harnessing of nuclear energy. Whatever the ultimate solution may turn out to be, we may rest assured that the fundamental basis for the energy will still be Einstein's simple equation, $E = mc^2$.

$E = mc^2$ _____

QUESTIONS

8–1. What is meant by the rest mass of an object?

8–2. The intrinsic mass of a fundamental particle, such as an electron, is often expressed in units of energy. Why is this justified?

8–3. If the mass of a moving particle is observed to be twice its rest mass, how much energy has been supplied to the particle, assuming it was initially in a state of rest? (Express your answer in terms of the rest mass, m_0, of the particle.)

Answer: $m_0 c^2$

8–4. If 1 gram of matter would convert to 20 trillion calories of energy, how long would it meet the calorie needs of a population of 2 billion persons? (Assume the daily calorie need of a person to be 2000 calories.)

Answer: 5 days

8–5. Suppose the speed of light is suddenly reduced by a factor of 2. Assuming nothing else has changed, what would be the change in the quantity of firewood to be burned if the initial rate of heat radiation is to be maintained?

Answer: must be increased fourfold

8–6. Energy conservation was a concept of great usefulness in pre-Einsteinian science, and still is in its relativistic version. Einstein was so committed to the principle of energy conservation that when the old version seemed to contradict relativity, he chose to broaden the concept of energy. Can you cite a similar example in a discipline familiar to you, where concepts are broadened so as to preserve a principle? Explain.

SUGGESTIONS FOR FURTHER READING

FORD, K. W. *Basic Physics*. Waltham, Mass.: Blaisdell Publishing Co., 1968. (Chapter 21.)

MARION, J. B. *A Universe of Physics*. New York: John Wiley & Sons, Inc., 1970. (Sections 7.5–7.7.)

9
Relativistic Momentum and Energy

Energy and momentum are two physical concepts that play an extremely important role in the study of motion. In Newtonian mechanics, they are related yet essentially separate concepts. In the theory of relativity, however, they emerge as absolutely inseparable, the link being of a fundamental nature. The present chapter will examine the concepts of energy and momentum in the light of relativity.

The Concept of Momentum

Momentum is a measure of the quantity of motion, and is defined as the product of the mass of an object and its velocity. That is, momentum = mass × velocity, or symbolically,

$$p = mv$$

where p stands for momentum.

In Newtonian mechanics, which is still valid where the velocity of moving objects is small compared to the velocity of light, the momentum of an object is directly proportional to its velocity, since the mass of an object is assumed to remain constant. In the theory of relativity, however, mass is not an invariant quantity, but increases with velocity. This fact may be explicitly displayed by writing the relativistic definition of momentum as:

$$p = v \left[\frac{m_0}{\sqrt{1 - \left(\frac{v^2}{c^2} \right)}} \right]$$

where m_0 denotes the rest mass.

Conservation of Momentum

The greatest utility of the concept of momentum is that momentum is a conserved quantity. A conserved quantity represents a number which remains the same no matter how many changes a particular process undergoes. The physicist calculates a conserved quantity at one moment, and if he calculates it again at a later time, after nature has completed all its changes, he will find that the conserved quantity is still the same as before. One, therefore, does not need to know what complicated sequence of changes a physical process goes through. In fact, the details of an interaction may be too complex, yet one may still be able to extract useful information about the interaction if one can determine what quantities are conserved. Such is the power and beauty of a conservation principle. Momentum is just such a conserved quantity.

As a way of illustrating the principle of momentum conservation, let us take the example of a rifle. The momentum of a rifle, when it is held by a stationary observer, is zero, because it has no velocity.

$$P_{\text{rifle}} = m_{\text{rifle}} \times 0 = 0$$

The conservation of momentum requires that when a rifle fires a bullet, the total momentum of the rifle-bullet system must remain unchanged, namely, zero. This would be so, regardless of how complicated the process of explosion might be. Now the bullet, of itself, has a momentum, because it is in motion. The rifle, then, must recoil, so that its recoiling momentum may just balance the forward momentum of the bullet.

Rocket propulsion is another example of momentum conservation. As the exhaust gas is ejected in one direction, the rocket recoils in the opposite direction, thus being propelled. However, there is no net change

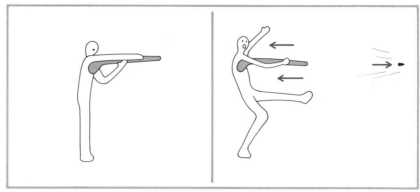

FIGURE 9–1.   Conservation of momentum requires that the rifle, upon firing, recoil.

in the momentum of the rocket-gas system, since the change in the rocket's momentum is balanced by the momentum of the exhaust gas.

Relativistic Energy

Mass represents the energy content of an object. A stationary object having a mass m_0 possesses an intrinsic energy of being, called the rest energy, which is given by the well-known Einstein relation, $E_0 = m_0c^2$ Rest energy means that if an object were ever completely converted into pure energy, the energy released would be exactly in the amount of the rest energy.

An object in motion possesses greater energy than when it is at rest. This is so, because in order to move an object, energy must be supplied to the object, and energy thus supplied becomes a part of the object's increased energy. But energy is equivalent to mass, and so the increased energy corresponds to an increase in mass. More precisely,

$$E = m \times c^2$$

where

$$m = \frac{m_0}{\sqrt{1 - \dfrac{v^2}{c^2}}}$$

We may call E the total energy of an object.

Relativistic Momentum and Energy_____

Kinetic energy is an energy associated with motion alone. An object at rest possesses no kinetic energy. Therefore, the total energy of an object is simply the sum of the rest energy and the kinetic energy.

$$\text{total energy} = \text{kinetic energy} + \text{rest energy}$$

or

$$E = K + E_0$$

The nonrelativistic expression for kinetic energy is

$$K_{\text{nonrel}} = \tfrac{1}{2} m_0 v^2$$

It is interesting to observe that this particular expression obtains from the relativistic definition of kinetic energy when the velocity involved is much less than the velocity of light. That is, for $v << c$ (see Question 9-4), $(E - E_0)$ reduces to $\tfrac{1}{2} m_0 v^2$. When the velocity of an object is very large, say 90 per cent of the speed of light, the nonrelativistic expression is not correct and cannot be used.

The Energy–Momentum Relation

The total energy of an object can always be calculated if its velocity is known. But the expression of the total energy in terms of momentum is also of great interest. A relationship between energy and momentum can be derived from the relativity of mass,

$$m = \frac{m_0}{\sqrt{1 - \dfrac{v^2}{c^2}}}$$

and takes, after some algebraic manipulations, the form of

$$E^2 = (pc)^2 + (m_0 c^2)^2$$

where $E = mc^2$ and $p = mv$.

The peculiar aspect of the relativistic energy–momentum relation is that the terms involved come as *squared* quantities. It turns out that this particular combination is mathematically inescapable. An implication of this peculiar combination is that in order to arrive at the energy of an object, one must take the square root of both sides of the energy–momentum equation, thus obtaining two values for the energy. The two possible values of the energy are:

$$E = + \sqrt{(pc)^2 + (m_0 c^2)^2}$$

or

$$E = - \sqrt{(pc)^2 + (m_0 c^2)^2}$$

The positive energy poses no difficulty, as it conforms to the observation that the rest energy (and rest mass) is always positive, and that the total energy increases with increasing momentum. The negative energy, on the other hand, is very strange: the higher the momentum, the more negative the energy becomes. This means that the faster an object moves, the lower its total energy becomes—a result that is contrary to common sense. If we consider an object whose momentum is zero, the peculiarity of the negative energy is even more apparent. The rest energy assumes a negative value, implying that the rest mass of the object must also be negative.

$$E_0 = -m_0 c^2$$

But we know that the notion of negative mass is unacceptable, because negative mass is not observable and therefore unphysical. One is tempted simply to discard the negative energy expression as meaningless. However, in the theory of relativity, the positive-energy solution does not really have a preferred claim over the negative-energy solution. To discard the negative energy, therefore, would be completely arbitrary and without justification.

What, then, is the meaning of the negative-energy solution? The negative-energy solution of the relativistic energy–momentum relation attains its true significance when it is interpreted in the light of quantum theory—a theory that unlocks the secrets of the microscopic world.

QUESTIONS

9–1. Consider a cheerleader and the earth as forming a system. When she is standing still, her momentum is zero. However, when she jumps upward, her momentum is no longer zero. If the total momentum of the cheerleader–earth system remains unchanged, how would you account for the change in her momentum?

Answer: Earth recoils.

Relativistic Momentum and Energy‗‗‗‗‗‗‗‗‗‗‗‗‗‗‗‗‗‗‗‗‗‗‗‗‗‗‗‗‗

9–2. In a head-on collision the cue ball comes to rest, and the struck billiard ball moves. Explain how momentum is conserved in such a collision on the pool table.

> *Answer:* Momentum of the cue ball is transferred to the struck billiard ball.

9–3. If a spaceship pilot wants to change the coasting speed of the spaceship in outer space, what must he do? Explain in terms of momentum conservation.

9–4. For $v \ll c$, the following relationship holds:

$$\frac{1}{\sqrt{1 - \left(\frac{v^2}{c^2}\right)}} \cong 1 + \tfrac{1}{2}\,\frac{v^2}{c^2}$$

Show that the difference between the total energy and the rest energy, $(E - E_0)$, reduces to $\tfrac{1}{2} m_0 v^2$.

9–5. Starting with the relativity of mass,

$$m = \frac{m_0}{\sqrt{1 - \left(\frac{v^2}{c^2}\right)}}$$

derive the energy–momentum relationship

$$E^2 = (pc)^2 + E_0{}^2$$

where $E = mc^2$ and $E_0 = m_0 c^2$.

9–6. Assume that an object, released to fall freely under gravity, falls downward because it has a *positive* mass.
   a. How would an object of negative mass fall under gravity?
   b. Argue why negative mass must be regarded as unphysical.

SUGGESTIONS FOR FURTHER READING

Cooper, L. N. *An Introduction to the Meaning and Structure of Physics,* short edition. New York: Harper & Row, Publishers, 1970. (Chapter 25.)

Smith, J. H. *Introduction to Special Relativity.* New York: W. A. Benjamin, Inc., 1965. (Chapter 9.)

Physics: The Fabric of Reality————————————————

10

The Emergence of Antimatter

The special theory of relativity has wrought a fundamental change in the physicist's picture of the world. Natural phenomena at very high speeds would be completely incomprehensible without relativity.

A Brief Remark About the Quantum

It was discovered at the turn of the century that the understanding of the microscopic world would require still another conceptual revolution. The occasion was the phenomenon of radiation emitted from incandescent objects. The well-known physics of Maxwell and Newton predicted that such objects would emit radiation of all frequencies, with the intensity of radiation increasing with the frequency. Thus an incandescent body would emit not only more ultraviolet radiation than infrared, but also an infinite amount of total radiation. This prediction was particularly significant, because as an inescapable consequence of the premises of classical physics, it put classical physics to a crucial test. It was, at the outset, obvious, however, that the prediction did not conform to reality. In fact, it was absurd that this theoretical result was referred to as the "ultraviolet catastrophe" of classical physics.

It seems that it is often the "catastrophic" results that jolt physicists to radically new ideas. Relativity was born out of the unexpected null result of the Michelson–Morley experiment. In the microscopic world it was the "ultraviolet catastrophe" that jolted the physicists to re-evaluate the very foundations of classical theoretical structures, such as those of Newton and Maxwell.

The fundamental hypothesis that successfully resolved the "ultraviolet" dilemma was the quantum theory proposed by Max Planck in 1900. Planck's quantum hypothesis, initially introduced to explain the phenomenon of incandescent radiation, assumed that changes take place in a discontinuous manner. Such an assumption clearly contradicted the accepted notion that physical changes occur in a continuous way. Nevertheless, it was just the revolutionary concept needed to open a new frontier to the microscopic world.

By the time P. A. M. Dirac emerged on the scene of theoretical physics in 1928, quantum theory had matured to a very successful theory. It was meeting up to the experimental tests of ever-growing refinement and sophistication. Dirac, however, was less than happy with the existing version of the quantum theory. His source of dissatisfaction was the observation that the theory, as it then stood, was inherently incompatible with the theory of relativity. His conviction that all physical theories should be consistent with the theory of relativity drove him to formulate a relativistic quantum theory in 1928.

The success of Dirac's theory was immediate. Among its best-known successes, the theory agreed completely with the observed spectrum of the hydrogen atom, and predicted the "magnetic strength" of the electron. Above all, it was a structure of great elegance and mathematical beauty.

Dirac's Theory of the Electron

In 1930 Dirac published the results of his relativistic quantum theory when it was applied to the problem of the free electron. By a free electron we mean an electron free of all external influences. By contrast, an atomic electron is bound to the atom and is, therefore, not free. But what is an electron? An electron is a physical entity distinguished by a certain set of intrinsic properties: a certain mass at rest, a negative electric charge, and so on. All electrons share the same intrinsic properties. Electrons are interesting, because they constitute one of the fundamental building blocks of the universe. The properties of the fundamental elements, for example, are predominantly determined by the behavior of the electrons in the atom.

The property of the electron that will concern us in this chapter is its mass. Mass is a relative concept, and so, when we refer to the mass of an electron, we must specify the electron's state of motion. When an electron is at rest, its mass is called the rest mass, m_0. The rest mass of an electron represents the intrinsic energy content of the electron and is often used interchangeably with the equivalent concept of rest energy. The rest energy of an electron is given by Einstein's mass-energy equivalence formula, $E = mc^2$, and has the numerical value of 0.51 MeV. (*MeV* is an abbreviation of million electron volt, and is a unit of energy commonly employed in atomic physics. One MeV is approximately equal to 4×10^{-12} calories.)

The rest energy of a particle clearly represents the absolute minimum energy that the particle can have and still retain its original particle character. Set in motion, the particle would simply have an increased energy. Specifically, the greater the velocity of a particle, the higher its energy would be. Accordingly the total energy of a particle would always be greater than, or at least equal to, its rest energy.

Thus, a free electron can assume any energy above its rest energy. This fact can be summarized in a so-called energy-level diagram. In such a diagram the energy is plotted vertically, with each allowed value of the energy indicated by a horizontal line. The energy-level diagram of a free electron would then consist of an infinite number of horizontal lines, with

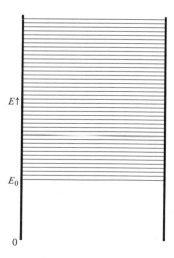

FIGURE 10–1. The energy-level diagram of a free electron, representing the observation that an electron can possess any energy equal to, or greater than, its rest energy.

The Emergence of Antimatter

the lowest line representing the rest energy. Such a conclusion seems consistent with observation.

Dirac's quantum theory, however, predicts that an electron can have not only a positive energy but also a negative energy, as a consequence of relativity. The positive energy of an electron is perfectly normal and presents no problems, as a positive mass is implied by it. In fact, every physical object in the entire universe is believed to have positive mass. The prediction that an electron may also have a negative mass, therefore, is a seemingly bizarre consequence. Either the negative-energy states must somehow be discarded in a justifiable manner or they must be interpreted so as to be physically meaningful.

In order to see if the negative-energy states could be discarded in a natural way, let us begin by constructing an energy-level diagram for Dirac's relativistic electron. On the positive energy side, the lowest energy an electron can assume is its rest energy, $E_0 = m_0 c^2$. But the electron is free to move about, and motion implies a higher mass, thus a higher energy. Since mass approaches infinity as the velocity approaches the speed of light, essentially an infinite number of positive energy states are available to the electron.

The negative-energy states are a mirror image of the positive-energy states, with the following important difference: An electron occupies the highest negative-energy state, $-m_0 c^2$, when it is not moving. Strange as it may seem, the faster a negative-energy electron moves, the lower its

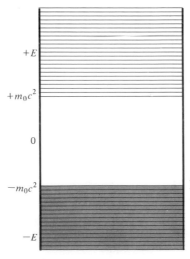

FIGURE 10–2. An energy-level diagram for Dirac's relativistic electron. The Dirac theory predicts negative as well as positive energy states for the electron: $E_+ \geqslant +m_0 c^2$ and $E_- \leqslant -m_0 c^2$.

energy becomes, thus losing energy as it speeds up. These rather peculiar negative-energy states are referred to as the Dirac Ocean.

Between the positive-energy states and the Dirac Ocean lies a region of forbidden energy states. No electron can exist there, because if it did, it would have a mass less than the rest mass of an electron. But such a particle would no longer be an electron.

Remarks About Transitions Between Energy States

Transitions between energy states are governed by the principle of energy conservation. Energy conservation implies that if an electron, say in a state E_0, absorbs an energy ΔE from an external energy source, the electron must be raised to a higher energy state E_h, which is defined by the sum of the initial energies. That is:

$$E_h = E_0 + \Delta E$$

The process of raising an electron to a higher energy state is called excitation. The mechanism of excitation that will concern us in this chapter is one via the absorption of a photon. (A photon is another name given to light.)

In the scheme of things, an excited electron always seeks to return to the lowest available energy state. In fact, it is the common tendency of all physical systems to seek the condition of greatest stability. The states of energy are defined in such a way that greatest stability is achieved when a

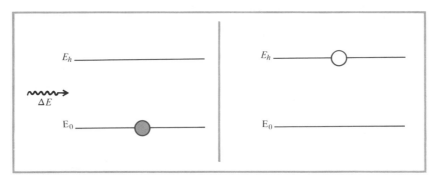

FIGURE 10–3.   Excitation by absorption of energy.

The Emergence of Antimatter_____

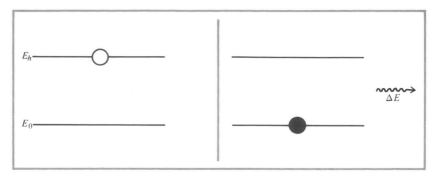

FigURE 10–4.   De-excitation or downward transition by emission of energy.

physical system attains the lowest energy state. The process of going from a higher to a lower energy state is termed a downward transition. Downward transitions are also governed by the principle of the conservation of energy, implying that when an electron goes from a higher energy state down to a lower energy state, it must give off the excess energy. The excess energy is usually radiated in the form of a photon.

The Dilemma of the Negative Energy States

Given an electron at rest, the question arises: Will it, seeing that there are lower energy levels available in the Dirac Ocean, make a downward transition?

In the classical physics of Newton and Maxwell, the negative-energy states pose no problem. An ordinary electron at rest, for example, cannot make a downward transition to the Dirac Ocean without first going through the forbidden energy region, because, in classical physics, all changes are assumed to take place smoothly and continuously. But to exist in an energy state in the forbidden region would imply a drastic change of identity: The particle in transition would no longer be an electron. A positive-energy electron is thus prevented from plunging into the oblivion of the Dirac Ocean. The electrons that happen to exist already in the Dirac Ocean, on the other hand, can simply be overlooked and discarded as unphysical in the sense that negative-mass particles are unobservable. They can be overlooked as long as no interaction exists between a positive-energy electron and the Dirac Ocean.

Physics: The Fabric of Reality

In a quantum theory, however, the negative-energy states pose a serious problem. The revolutionary aspect of the quantum theory is that changes on an atomic scale do not take place in a continuous manner, but rather in discontinuous steps. A quantum system going from an initial state to a final state, for instance, must be viewed as disappearing in the initial state to appear later in the final state. What happens in between is an irrelevant question in the new theory.

A quantum electron, then, can legitimately make a quantum jump over the forbidden energy region, and splash down in the Dirac Ocean. In fact, there is nothing in the Dirac theory that would prevent an ordinary electron from making the initial transition to a negative-energy state, and then subsequent transitions to lower and lower negative energy states, radiating more and more energy in the process. The implication is that all ordinary electrons would ultimately cascade down to the Dirac Ocean, and thus into utter oblivion. The theory further predicts that it would take an electron in a hydrogen atom, for example, only about 10^{-10} seconds to descend into the Dirac Ocean, never to return again! This is clearly nonsense. Hydrogen atoms are presumably as old as the universe itself, and there is at present an abundance of ordinary hydrogen atoms, consisting of ordinary electrons.

One is thus confronted with a situation where a new theory, which is brilliantly successful in explaining observed atomic spectra, predicts also magnificent absurdities. One must wonder if Dirac's relativistic quantum theory, despite its delightful elegance, has fallen by the wayside as another catastrophe in man's desperate search to know. To forge ahead amid adversities, however, is one of the noblest qualities of the human spirit.

Dirac's Resolution

In 1925 an important principle was discovered about a peculiar behavior of electrons. From a careful analysis of atomic spectra, W. Pauli deduced that no two electrons could occupy the same energy state. Once an electron exists in a particular state, it excludes all the other electrons from sharing the same state. This observation is called the exclusion principle. As no electron is exempt from the exclusion principle, Dirac's electrons must also obey the Pauli exclusion principle, regardless of whether they are in the positive-energy states or in the negative-energy states.

The Emergence of Antimatter_____

So, in 1930, Dirac forged ahead with his theory, proposing that the ocean of negative-energy states be regarded as existing already filled to capacity with the unphysical negative-mass electrons. The exclusion principle would then dictate that the Dirac Ocean could accommodate no electrons from a positive-energy state. Ordinary electrons would thus be compelled to remain in the positive-energy states, as if supported by a filled Dirac Ocean. Dirac's assumption of a filled negative-energy sea was, to be sure, a daring, if not a far-fetched, idea, but it did resolve the problem of the catastrophic electron descent.

Holes in the Dirac Ocean

The Dirac Ocean has an infinite negative mass as well as an infinite negative charge, since it would require an infinite number of electrons to fill it. It is, in a sense, everywhere. Why are we, then, completely unaware of it? G. Gamow has offered the analogy of a dolphin. Although it is surrounded by water on all sides, a dolphin may be unaware that it exists in a physical medium. The medium simply constitutes a part of its normal existence. On the other hand, local disturbances in the surroundings, such as air bubbles, would be observable to the dolphin.

Perhaps local disturbances in the Dirac Ocean may also be observable.

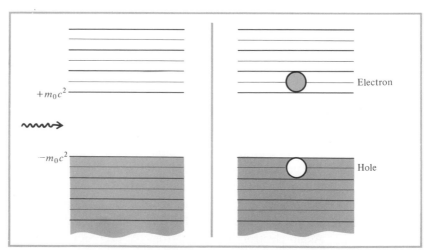

FIGURE 10–5.   A photon leading to the creation of an electron and a "hole" in the Dirac Ocean.

Let us grant that ordinary electrons are prevented from descending into a filled Dirac Ocean. But is there anything that would prevent an electron from jumping out of the Dirac Ocean?

Dirac argued that a negative-mass electron in the Dirac Ocean could indeed be excited to a positive-energy state by absorbing radiation from an external source. Upon such excitation, the electron would be observable as an ordinary electron, since it would now have a positive mass. But, at the same time, the raised electron would leave behind a "hole" in the Dirac Ocean.

A hole in the Dirac Ocean would, first of all, have a positive mass, as it would represent an absence of negative mass. It would, secondly, carry a positive charge, because the absence of negative charge must appear as a positive charge. As an entity with such observable properties as positive mass and positive charge, Dirac's hole would then be as real as an ordinary electron. Thus a photon absorption by the Dirac Ocean would result in the creation of a pair of particles—namely, an ordinary electron and its hole counterpart. Dirac further showed that such a pair, aside from their opposite charge, would behave in a remarkably similar manner in their interactions with the rest of the universe.

Between the electron and its hole counterpart, however, there exists a fundamental difference. A hole owes its total existence solely to the departed electron. Such a hole would then imply that the Dirac Ocean is no longer completely filled, thus making an empty energy state available to an

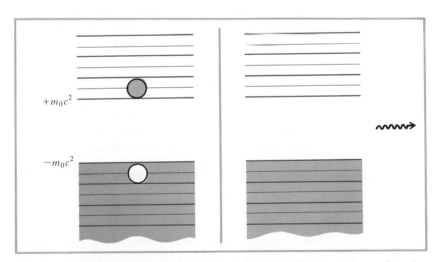

$+m_0c^2$

$-m_0c^2$

FIGURE 10–6.   Annihilation of an electron and an antielectron (or positron).

The Emergence of Antimatter

ordinary electron. Dirac's theory predicts that a hole at rest would last in the neighborhood of about 10^{-8} seconds before suffering annihilation. Annihilation would occur when an ordinary electron makes a downward transition and fills the empty negative-energy state. The only remnant in such an annihilation process would be the photon given off by the descending electron. The ultimate fate of a hole in the Dirac Ocean would thus be annihilation. Dirac called the hole an antielectron.

Experimental Confirmation

In 1932, just two years after Dirac's prediction of the antielectron, C. Anderson identified a unique track in his cloud chamber photographs of cosmic rays. He interpreted the track to have been made by a particle carrying a positive charge but having a mass comparable to that of an electron. The particle was thus identified to be the antielectron predicted by Dirac.

It has, since, been established that every fundamental particle has an antimatter counterpart. The preponderance of matter in our part of the universe is attributed to the fact that when matter encounters antimatter, annihilation results. In order to be observed in our world, antimatter must thus be created.

The fleeting existence of antimatter is a beautiful example of Einstein's mass–energy equivalence. The energy released upon annihilation of antimatter comes from the total conversion of the masses involved. In fact, the annihilation of matter and antimatter is the only way whereby mass may be completed converted into energy.

The universe that would be comprehensible within the union of relativity and quantum theory must necessarily consist of matter and antimatter. The fact that the totally unsuspected existence of antimatter was first predicted on purely theoretical grounds is a tribute to the power of pure thought, and to the faith that a theory describing a self-consistent universe must itself be self-consistent.

QUESTIONS

10–1. Suppose there exists an intelligent population somewhere in the universe. The members of the population slide from place to place. Their photograph of earth reveals "foot prints" on snow. If they interpret these "foot prints" to have been made by earth beings, what would they hypothesize about the manner in which the earth beings move? Explain.

10–2. Recapitulate the sequence of steps that Dirac followed in resolving the problem of the catastrophic electron descent into the Dirac Ocean. Which step do you think was most daring?

10–3. What minimum energy must a photon have in order to excite an electron in the Dirac Ocean to a positive energy state. (Assume the rest energy of the electron to be 0.51 MeV.)

Answer: 1.02 MeV

10–4. In the Dirac theory, the existence of an ordinary electron must be accompanied by a filled Dirac Ocean. The ocean, to be filled, requires an infinite number of unphysical negative-mass electrons.
a. What would be the mass of the Dirac Ocean?
b. What would be the electric charge of the Dirac Ocean? Recall that all electrons carry negative charge.
c. Under what conditions would the Dirac Ocean correspond to a vacuum?

10–5. An antielectron is also referred to as a positron, because it carries a positive charge in contrast to the electron's negative charge. When an electron and a positron, both at rest, annihilate, a pair of photons are produced. What is the energy carried by each photon?

Answer: 0:51 MeV

10–6. A proton is a positively charged particle whose mass is some 2000 times as great as the mass of the positron. Why does a proton not annihilate an electron, whereas a positron does?

10–7. A proton would be annihilated by an antiproton. What would be the charge of an antiproton? Explain in terms of a Dirac Ocean of unphysical protons.

Answer: negative

10–8. Antimatter cannot be created without creating matter at the same time. Speculate as to the possible existence of an antiuniverse.

The Emergence of Antimatter_____

SUGGESTIONS FOR FURTHER READING

Burbidge, G., and F. Hoyle. "Anti-matter," *Scientific American* (April 1968).

Cooper, L. N. *An Introduction to the Meaning and Structure of Physics,* short edition. New York: Harper & Row, Publishers, 1970. (Chapter 40.)

Gamow, G. "The Exclusion Principle," *Scientific American* (July 1959).

11
The Union of Space and Time

Man has always longed for absolutes. To be tied to something immovable, to know something immutable is to feel secure in a constantly and often violently changing world. The theory of relativity has, in a sense, been a violent revolution. The age-old absolutes of space and time—the corner stones of the familiar Newtonian world view—have now been demolished along with a host of other absolutes. To speak of absolutes again in the new world view would thus seem to be heresy, but the wonder of relativity is that it does not reject all absolutes. Rather, it replaces the familiar ones with new ones of its own, and in so doing probes deeper into the physical reality of the universe.

The Relativistic Invariant—A Definition

A relativistic invariant may be defined as a physical quantity that remains the same to all inertial observers. The speed of light is an invariant, and in a broader sense, the physical laws are also invariant as having the same mathematical structure to all observers. But such well-known concepts as space and time, or momentum and energy, are not relativistic invariants. Rather, they are relative concepts, assuming different values to different observers. It is out of such relative concepts, however, that relativity constructs new invariants to serve as relativistic absolutes.

An Energy–Momentum Invariant

A relativistic invariant for energy and momentum may be defined by uniting energy and momentum in the following way:

$$E^2 - (pc)^2 = E_0{}^2$$

where E and p denote the energy and momentum of a particle, respectively, as measured by an inertial observer. E_0 is then referred to as an energy–momentum invariant.

As a way of establishing that E_0, indeed, represents an invariant quantity, let us examine the preceding equation from the viewpoint of a particular observer. Suppose a particle has a zero velocity according to this observer, and thus a zero momentum. The total energy of the particle would then simply be its rest energy, m_0c^2.

$$E^2 - 0 = E_0{}^2$$

or

$$m_0c^2 = E_0{}^2$$

But the rest energy of a particle is the total energy of the particle in a frame of reference attached to the particle itself, and all observers would agree that such is indeed the case. Thus the rest energy of a particle, E_0, represents an invariant quantity.

The existence of an energy-momentum invariant implies that energy and momentum are linked in an inseparable way. Momentum, on one hand, is a three-dimensional concept, as its specification requires a set of three numbers, namely, the $x, y,$ and z components of momentum. Energy on the other hand, is a one-dimensional concept. The union of energy and momentum then constitutes a four-dimensional concept, and is often referred to simply as a "four-momentum."

A Space–Time Invariant

A relativistic invariant exists also for space and time. A space–time invariant, which achieves such a union, may be defined[1] as the sum of

[1] R. H. March, *Physics for Poets* (New York: McGraw-Hill, 1970), p. 145.

the squares of space and time intervals. Denoting the space interval by s and the time interval by t, the invariant I_0 is given by

$$s^2 + (ct)^2 = I_0^2$$

Note how structurally similar the space–time relation is to the energy–momentum relation. The reason why momentum and energy are referred to as the space and time components of "four-momentum" is now obvious.

That this particular combination of space and time leads to a quantity that is the same to all observers can readily be demonstrated. Introducing two observers, Joe and Moe, again, let us analyze two events that are simultaneous to Moe but not to Joe. Suppose that the spaceship observer Moe observes two light receivers, separated by a spatial interval, s_{moe}, to trigger at the same instant of time. That is, the time interval between the two events is zero, or $t_{\mathrm{moe}} = 0$, so that the space–time relation takes the form:

$$s_{\mathrm{moe}}^2 = I_0^2$$

Observing the same two events taking place in the moving spaceship, the ground-based observer Joe would conclude that the spatial interval between the two light receivers in the spaceship is shorter than that measured by Moe according to the length-contraction equation:

$$s_{\mathrm{joe}} = s_{\mathrm{moe}} \sqrt{1 - \left(\frac{v^2}{c^2}\right)}$$

Joe would further conclude that the events are not simultaneous, and deduce from the relativity of clock synchronization that the time interval between the events must be

$$t_{\mathrm{joe}} = \frac{s_{\mathrm{moe}}\, v}{c^2}$$

which is adapted from the now-familiar fact that the "rear" clock is ahead of the "front" clock by a time interval, $L_0 v / c^2$. The space–time relation then reduces to

$$s_{\mathrm{joe}}^2 + (ct_{\mathrm{joe}})^2 = I_0^2$$

or, after a bit of algebra,

$$s_{\mathrm{moe}}^2 = I_0^2$$

But s_{moe}^2 is precisely the value that Moe deduced the invariant quantity

$I_0{}^2$ to be. Thus, both Joe and Moe agree on the value of I_0. A space–time invariant, therefore, does exist.

An important and subtle feature of the space–time union is that space and time are interchangeable—a strange and startling idea. Returning to our thought experiment, let us note that Moe observes only a space interval, the time interval being zero. Joe, on the other hand, observes both space and time intervals. However, the space interval can be made to approach zero by increasing the speed of the spaceship to near the speed of light. Such a drastic space contraction would then make the time interval dominate in Joe's observation. Joe would thus observe that the two events in the spaceship take place nearly on top of each other spatially but with an unambiguous time delay. Contrast this with Moe's observation that the same two events take place spatially separated but with no time delay between them. A space interval to one observer may thus appear as a time interval to another.

In the theory of relativity, then, space and time cannot be viewed as separate concepts, each existing independent of the other. The correct way is to view space and time as inextricably united, with three dimensions of space and one dimension of time merging to form a four-dimensional space–time.

The Minkowski Space–Time Graph

The three-dimensional Cartesian coordinate system may now be extended to a four-dimensional coordinate system. Such a four-dimensional space–time coordinate system is called a Minkowski graph, and consists of vertical and horizontal axes. The vertical axis represents the dimension of time, and the horizontal axis represents the three dimensions of space. If it is difficult enough to visualize four dimensions, it is no easier trying to represent them in the two dimensions of a flat page!

A point in the Minkowski space–time graph is called an event, specifying the "where" and "when" of an object. The history of an object is traced by a space–time line called a world line. Physics then becomes a study of the world lines that nature sketches on the canvas of space–time.

A closer look at the world line raises an interesting question: Is it possible for the arrow of a world line to point downward (backward in time), and if so, what would be its significance? The question is by no

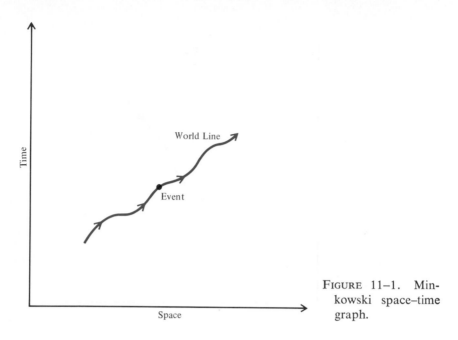

FIGURE 11–1. Minkowski space–time graph.

means a trivial one. After all, an object is free to move in any direction in space—right and left, or up and down. Furthermore, space and time are interchangeable in the sense that a space interval to one observer may appear as a time interval to another. It seems, then, that a particle should be allowed to move both forward and backward in time, unless relativity, in another context, provides a reason for a disparity between space and time. But no such disparity exists in the theory of relativity! A particle must be allowed to exist in a backward as well as forward direction of time. The poetic image of time as "flowing as a river," thus, becomes no longer tenable. Rather, the relativist must now envision an object, such as an electron, swimming upward or downward with equal facility in the stationary river of time.

Wheeler's Vision of a Relativistic Electron

In his 1965 Nobel lecture, entitled "The Development of the Space–Time View of Quantum Electrodynamics," [2] Professor R. P. Feynman

[2] R. P. Feynman, *Physics Today,* 19, 8 (August 1966), pp. 31–44.

The Union of Space and Time_____

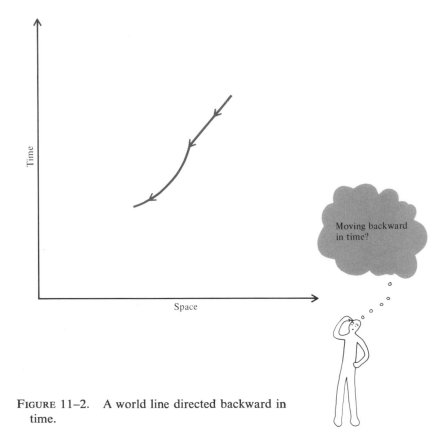

FIGURE 11–2.   A world line directed backward in time.

relates a telephone conversation he had had with Professor J. A. Wheeler during his graduate student days at Princeton:

"Feynman, I know why all electrons have the same charge and the same mass," said Wheeler.
"Why?"
"Because they are all the same electron."

Wheeler explains his new vision of the electron in terms of its world line in space–time. He envisions the entire cosmos to be covered with a rich tapestry woven of the electron's world lines—as a result of, say, a single electron shuttling back and forth on the loom of time. Looking through a time slot called "now," we would then see many, many world lines representing the electrons observed in the universe. What we observe as electrons would be nothing more than the cross sections of the knotted path of this single electron dancing in the vastness of a space–time continuum.

Physics: The Fabric of Reality————————————————

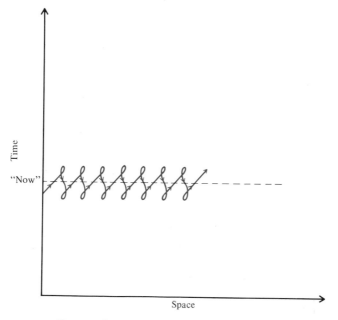

FIGURE 11–3.
Wheeler's vision of an electron dance on the stage of space–time. (Adapted from M. Gardner.)

But an electron cannot dance freely on the stage of space–time without stepping backward in time. To each world line that cuts through the time slots in the forward direction, there must be another world line in the backward direction of time. The world lines directed forward in time represent the ordinary electrons, but what could be the significance of those world lines pointing in the backward direction of time?

A clue lies in the behavior of the electron in the presence of an electromagnetic field. The electromagnetic interaction is such that whatever a positive electric charge does, a negative electric charge does exactly the opposite. So if an electron is attracted into the past (backward in time) by an electromagnetic field, a positron—the positively charged antimatter counterpart of the electron—would be propelled into the future (forward in time) by the same electromagnetic field. A positron moving forward in time, then, may be regarded as an electron moving backward in time.

This is a fantastic idea—that an electron dancing across the "now" line in space–time may exist as a positron as often as an electron. Furthermore, there would be as many electrons as positrons in the universe, implying a symmetry between electrons and positrons—a feature that physicists always find pleasant.

Wheeler was not really proposing a serious theory, or, perhaps, he was. Anyway, the idea had hidden in it a gem of truth for Feynman.

The Union of Space and Time

101

The Space–Time View of Positrons

In the same lecture, Feynman goes on to say,

> I did not take the idea that all electrons were the same one from him as seriously as the observation that positrons could simply be represented as electrons going from the future to the past in a back section of their world lines. That, I stole!

The seed of an idea, incredible as it sounds, fell in the soil of a fertile mind.

Recall that what led Dirac to the existence of the antielectron was the negative sign of the total energy expression:

$$E = \pm \sqrt{(pc)^2 + E_0{}^2}$$

The space–time equation is also a quadratic equation, and when one solves for time, one finds that the solution may assume either a positive or a negative value.

$$t = \pm \frac{\sqrt{I_0{}^2 - s^2}}{c}$$

FIGURE 11–4. The reversal of the direction of an electron world line by catastrophic emission of a photon.

The dramatic discovery of Feynman was that a positron may be regarded as a negative-energy electron moving backward in time.

In Feynman's space–time view, electron and positron—and, in general, matter and antimatter—need no longer be viewed as separate concepts. Matter and antimatter are really two aspects of the same thing—the only difference being the sense of time and energy. Furthermore, if we detach ourselves from our local confinement and see all of space and time on a single canvas, we may, indeed, see only one electron world line zigzagging across the "now" line.

But what causes an electron suddenly to reverse its direction in time at certain points in space–time? At least this much is clear—that apart from an actual physical process, such time reversals could not occur. What Feynman showed is that a vertex in a space–time graph does not occur all by itself in a vacuum, but must necessarily involve the presence of another object called a photon. The actual physical process may be represented in a space–time graph in the following way. The electron reverses its direction in time by catastrophically emitting a photon at the vertex.

To interpret such a process, let us scan the diagram upward through our time slot. We would see, to begin with, two world lines—one in the forward direction of time, representing an actual electron, and the other

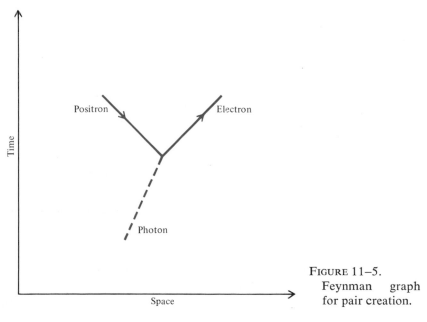

FIGURE 11–5.
Feynman graph
for pair creation.

The Union of Space and Time

in the backward direction of time, representing an observable positron. The two world lines merge at the vertex, and a single world line representing a photon emerges. In terms of particles, an electron encounters a positron at a point in space–time, and they are annihilated producing a photon.

$$\text{electron} + \text{positron} \rightarrow \text{photon}$$

An electron moving backward in time may also reverse its direction in time by absorbing a photon. In terms of particles, this process represents the creation of an electron–positron pair.

$$\text{photon} \rightarrow \text{electron} + \text{positron}$$

Thus, when a positron is created, it is always accompanied by the simultaneous creation of an electron.

The space–time view of antimatter has proven to be a fruitful approach to the understanding of the universe. It has also opened up a host of interesting questions. In a more speculative vein, one could ask if man, too, could reverse his direction in time? This much seems certain. If he is going to reverse his time sometime in the future, his antiman counterpart must have already been there, and must currently be on a backward journey in time at some distant antiplace.

QUESTIONS

11–1. The relativity of mass,

$$m = \frac{m_0}{\sqrt{1 - \dfrac{v^2}{c^2}}}$$

leads to $E^2 = (pc)^2 + (m_0 c^2)^2$.

a. Write this expression in a primed and an unprimed coordinate system.

b. Show that in both coordinate systems the energy–momentum invariant E_0, as defined in the text, comes out to be $m_0 c^2$.

11–2. Suppose that the East and the West sign a peace treaty. To celebrate the occasion both parties agree to raise each other's flag at the *same* moment in their respective capitals. How would the flag-raising ceremonies appear to a fast-moving cosmic observer?

11–3. An interval in space–time is considered more fundamental in nature than an interval in time or in space. Explain.

11–4. In the foreward to *Physics in the Twentieth Century* (MIT Press, 1972) by V. F. Weisskopf, Professor Bethe writes, "While popular belief accuses science of having made everything relative, Weisskopf points out that relativity theory is really a theory of the absolute, of the quantities that remain unchanged from any standpoint (invariants)." Comment on this statement.

11–5. a. Draw the world line of a parked car in a space–time graph.
b. Draw the world line for an encounter between man and his antiman counterpart.

11–6. Comment on the remark that space–time constitutes a four-dimensional stage upon which nature performs.

11–7. Although the theory of relativity of itself does not predict the existence of matter, the quantum theory, apart from union with relativity, would not predict the existence of matter. What aspects of the relativity theory suggest the possibility of antimatter?

11–8. Speculate on the existence of an antiuniverse. What arguments can you offer in defense or in refutation of its existence?

SUGGESTIONS FOR FURTHER READING

FEYNMAN, R. P. "The Development of the Space–Time View of Quantum Electrodynamics," *Physics Today* (August 1966).

FORD, K. W. *Basic Physics*. Waltham, Mass.: Blaisdell Publishing 1968. (Section 27.2.)

GARDNER, M. "Can Time Go Backward?" *Scientific American* (January 1967).

MARCH, R. H. *Physics for Poets*. New York: McGraw-Hill Book Company, 1970. (Chapter 11.)

The Union of Space and Time⸺⸺⸺⸺⸺⸺⸺⸺⸺⸺⸺⸺⸺⸺

12
The General Theory of Relativity

"There is such a thing as a passionate desire to understand, just as there is a passionate love for music. Without it, there would be no natural science, and no mathematics." [1] So wrote Einstein in 1950, as he reflected on his relativistic theory of gravitation.

The General Principle of Relativity

Although the special theory of relativity grew in favor with every new observation, Einstein was not completely satisfied with it. He considered the special role of the inertial frames of references as a serious weakness of the theory and thought that the physical laws valid only to inertial frames could not be expected to be of general validity. He thus sought to generalize the principle of relativity so that all frames of reference, both inertial and noninertial, would be completely equivalent. The result was his proposal of a general principle of relativity, which may be stated as: *The laws of physics must be so formulated as to be invariant to all observers, inertial as well as noninertial.*

The theory that rests upon the foundation of the general principle of

[1] A. Vallentin, *The Drama of Albert Einstein* (Garden City, N.Y.: Doubleday, 1954).

relativity is the general theory of relativity. The task of the new theory is essentially to formulate the laws of physics so as to be invariant to all frames of reference. The laws of electromagnetism, for instance, are valid only to inertial frames, and if they are to conform to the requirement of the general principle of relativity, they must be reformulated. Einstein directed a great deal of effort to a reformulation of electromagnetism, but did not succeed in producing a general relativistic theory of electromagnetic phenomena. The laws of gravitation, on the other hand, were successfully formulated so as to be invariant to all frames of reference, leading to new insights not achieved in Newton's theory of gravitation. Einstein's general theory of relativity is, for this reason, essentially a relativistic theory of gravitation. It is conceivable that the general theory of relativity may eventually be broadened to incorporate the phenomena of electromagnetism, thus bringing about another synthesis of knowledge.

Noninertial Observers

A noninertial observer is one in a state of accelerated motion. Acceleration is defined as the rate of change of velocity.

$$\text{acceleration} = \frac{\text{change of velocity}}{\text{time elapsed during change}}$$

When it reaches a speed of, say, 60 m.p.h. starting from rest, an automobile is said to have accelerated. The magnitude of the acceleration depends on the length of time involved in attaining the final speed. The quicker the final speed is reached, the greater the acceleration is.

Acceleration can be uniform or nonuniform. By uniform acceleration we mean that the change in velocity is the same for all equal intervals of time. An automobile accelerating uniformly, for example, would take the same time in going from zero to 30 m.p.h. as from 30 to 60 m.p.h. We shall not consider nonuniform acceleration in this discussion.

Accelerated motion is uniquely detectable, contrary to the nondetectability of uniform motion. Consider a spaceship in outer space, undergoing sudden acceleration. As long as it is coasting at a constant velocity, every object in the spaceship would remain suspended exactly where it is placed. But once the spaceship is accelerated, for instance, upward, all the freely floating objects would fall downward to the floor. These effects would be unmistakable. Furthermore, observations of these falling objects would enable the pilot to determine the magnitude and direction of the spaceship's acceleration.

Rotation also constitutes an accelerated motion. The water surface of a pail in an inertial frame is flat, but set in rotational motion the water surface would form a concave shape. The difference is, again, unmistakable, and therefore, a state of rotational motion can also be uniquely determined.

The Principle of Equivalence

The special theory of relativity rests on the postulate of the equivalence of all inertial observers. In the formulation of the general theory of relativity, on the other hand, Einstein is primarily concerned with accelerated motion. Consider again an accelerating spaceship in outer space where there are no external physical influences. Let the pilot release two objects of different shape, size, and mass in the middle of the ship. The pilot would observe that they fall to the floor exactly at the same rate.

This thought experiment in the accelerating spaceship is reminiscent of the legend of Galileo and the Leaning Tower of Pisa. What Galileo

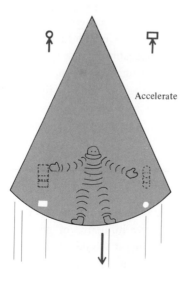

FIGURE 12–1. Spaceship in accelerated motion.

demonstrated there was that when objects fall because of gravity, they fall at the same rate, in the absence of air resistance. Einstein's genius was to recognize this familiar phenomenon as a fundamental link between gravity and accelerated motion.

Another familiar observation suggesting a possible link between gravity and acceleration is that gravity may be canceled by an acceleration. An astronaut orbiting the earth experiences "weightlessness." He is weightless because the earth's gravitational pull on him is exactly canceled by the effect of acceleration of the orbiting satellite.

Observations such as these provided Einstein with a hint for a postulate which would eventually lead to a successful theory of gravitation. He recognized this apparent link between gravitation and acceleration to be of a fundamental nature, and thus raised it to the status of a postulate. It may be stated: *It is impossible to determine whether one is in a state of accelerated motion or in the presence of gravitation.* Or it may be stated, equivalently: *The effects of accelerated motion are indistinguishable from those of gravity.* This postulate is called the principle of equivalence and serves as the foundation for Einstein's general theory of relativity.

An Experimental Test

An experimental test of the equivalence principle would consist in comparing the results observed in an accelerated spaceship with those observed under the influence of gravity. Consider a light beam emitted from the ceiling of an accelerating spaceship. If the spaceship is moving at a certain velocity at the moment of light emission, the spaceship would be moving at a higher velocity at the moment of light absorption by the detector fixed on the floor of the spaceship. The accelerating motion of the spaceship would then imply a relative motion between the emitter at the moment of light emission and the detector at the moment of light absorption. Such a relative motion would produce a Doppler shift, thus rendering the frequency of the light at the detector higher than the frequency of the light at the emitter.

The principle of equivalence would then imply that a similar effect must also be detectable under the influence of gravity. A gravitational analogue of the spaceship experiment would place a light-emitter at the top of a tower and a detector on the ground. Such an experiment was performed by R. V. Pound and G. A. Rebka in 1960. They placed a light source at the top of a tower 76 feet high and measured the frequency of the light on the ground. Einstein's theory predicted that the fractional shift in the frequency of light at the two positions would be 2.3×10^{-15}. Pound and Rebka observed a fractional frequency shift of 2.4×10^{-15}, in agreement with the theoretical prediction well within the experimental accuracy.

Curved Space

A thought experiment enables us to follow the path of a light beam in an accelerating spaceship. For the sake of convenience, let us assume that a light beam is flashed horizontally, and employ for a detector a series of parallel fluorescent plates that are stacked vertically and equidistantly from one another. The light would, then, make flashes in the plates as it travels across, and the line linking all the flashes would yield the path of the light beam.

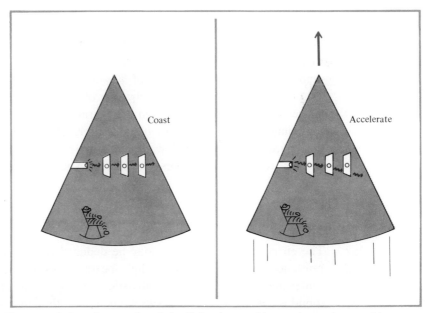

FIGURE 12–2.   A curved path for light observed in accelerated spaceship.

The General Theory of Relativity

As the light propagates across, the fluorescent plates are continuously displaced upward because of the acceleration of the spaceship. If the first flash occurs at the midpoint of the first plate, the second flash would occur slightly below the midpoint of the second plate, and the third flash would occur still further below. The net effect is that the line connecting all the fluorescent flashes would appear to be curved to the observer in the spaceship. He would then conclude that light falls toward the floor of the accelerating spaceship, thus following a curved path.

But the observer in the spaceship cannot determine whether he is in a state of accelerated motion or under the influence of gravity. It would therefore be equally correct to say that light falls in a gravitational field, and that its path is bent by gravity.

The idea that light follows a curved path is a strange one. One usually thinks of the path of light as always being a straight line, because a familiar definition of a straight line is that it is the shortest distance between two points, and one takes it for granted that light always takes the shortest path. But a straight line constitutes the shortest path only in flat space. On the surface of the earth the shortest path between two points is not a straight line. A jetliner, for example, would follow a curved path over the North Pole if it were to take the shortest route between New York and Peking.

Einstein assumed that light always follows the shortest path in space. Therefore, he interpreted the bending of light by gravity as suggesting that space itself is curved. The curvature of space was regarded as a manifestation of the presence of physical matter, a greater curvature resulting from a greater concentration of matter. The notion of a flat space thus became untenable in a general description of the universe, obtaining only in the absence of all matter. The Euclidean geometry of flat space was thus found wanting. A correct picture of the physical universe required a new geometry, namely, a geometry of curved space.

"Warped" Time

The effect of gravity on time may be deduced from a rotating frame of reference. Suppose two similar light-beam clocks are each placed at the center and at the edge of a relativistic merry-go-round. If the merry-go-round is not rotating, both clocks would tick at the same rate to a ground-observer. Set in rotating motion, however, the clock on the edge would advance at a slower rate than the clock at the center. The reason is that the clock on the edge is in motion relative to the ground-based observer, whereas the clock at the center is not.

Let us now observe the two clocks from the viewpoint of an observer at the center of the merry-go-round. As there is no relative motion between the observers, the merry-go-round observer would agree with the ground-based observer—that the clock at the edge advances slower than the clock at the center. However, neither clock is moving with respect to his frame of reference. The only difference in the states of motion of the clocks would be that the clock at the edge is in accelerated motion, whereas the clock at the center is at rest. The slower rate of the edge clock must then be attributed entirely to the effect of the acceleration that the clock experiences. If the merry-go-round is rotated at a faster rate, the clock on the edge would experience a greater acceleration, and would thus advance still slower.

According to the principle of equivalence, the effects of accelerated motion must also occur in the presence of gravity. Gravity, therefore, must also slow down the clock rate. As a clock is moved from a place of weaker gravity to a place of stronger gravity, its rate would decrease. A clock would advance at different rates at different points in the universe, But, then, so must time, for time is what a clock measures. Gravity is thus said to "warp" time. Time warp is a truly profound discovery about gravity and time.

Observational Evidence

Almost all the predictions of Einstein's theory are identical to those of Newton's theory of gravitation. What distinguishes one theory from the other is a handful of extremely small gravitational effects which Einstein's theory predicts but which Newton's theory does not. Experimental tests of the general theory of relativity, therefore, consist in observing these small effects.

Bending of Light by Gravity

The gravity of the earth is too weak for the bending of light to be measurable. For all practical purposes light does travel in a straight line on earth, and space in the neighborhood of the earth is therefore flat. But the gravitational strength of the sun is much stronger. It was Einstein himself who suggested an experimental test for space curvature by predicting that a ray of starlight passing close to the sun would be bent by an extremely small yet measurable amount, namely 1.75 seconds of arc (4.7×10^{-4} degrees of angle).

The first experiment was performed during a total solar eclipse in 1919. An international expedition photographed the apparent positions of stars whose light passed near the sun in approaching the earth. From an independent knowledge of the positions of the same stars when the sun was elsewhere, they deduced the amount of the bending of light caused by the sun's gravity, and arrived at a deflection of light in general agreement with Einstein's predicted value. However, the precision required to confirm the theory unequivocally has not yet been achieved.

The Effect of Gravity on Time

Another prediction of Einstein's theory is that a clock would advance at different rates depending on the strength of gravity. For instance, time would advance faster at the top of a skyscraper than at the ground level. A hostess on the top floor of the Empire State Building would—at least in principle—age faster that her twin sister working on the ground floor.

A cosmological implication of this effect is that atoms would vibrate at

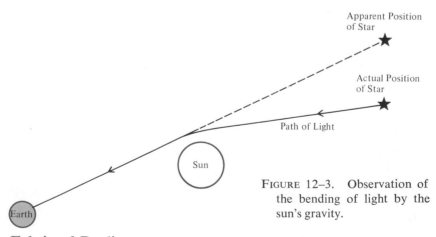

FIGURE 12–3. Observation of the bending of light by the sun's gravity.

Physics: The Fabric of Reality

a slower rate in the sun or in stars than on earth. Red light emitted by a solar atom would, for instance, vibrate at a slightly slower rate than the red light emitted by a similar atom on earth, thus appearing slightly less red. Similarly, blue light from the sun would appear slightly less blue to an earth-bound observer. This effect is called a gravitational red shift. An observation of the gravitational red shift in the sun was made by J. Brault in 1962. The experimental value was found to agree with the predicted red shift within 5 per cent.

The Orbital Motion of Mercury

Newton's theory of gravitation predicts an elliptical path for all planets. The point of the orbit closest to the sun is called the perihelion. Newton's theory further predicts that the perihelion of each planet does not return to the same place (relative to the fixed stars). This is due to the mutual gravitational interactions among the planets themselves. However, because these interactions are vastly smaller than the interaction between the sun and the planet, the advance of the perihelion is an extremely small effect. Newton's prediction, however, is not in complete agreement with the observed advance of the perihelion of the innermost planet Mercury. Newton's theory, taking into account the effects of all other planets on Mercury, predicts an advance of 500 seconds of arc per century for the perihelion of Mercury relative to the fixed stars. The dis-

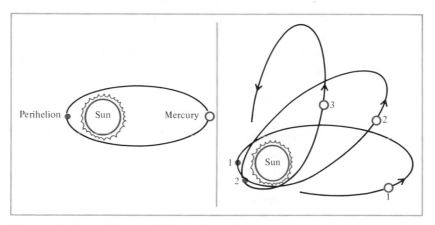

FIGURE 12–4.   Advance of Mercury's perihelion.

The General Theory of Relativity

crepancy between observation and Newton's theory is something like 43 seconds of arc per century, but negligible as this may seem, it is sufficient to point to a fundamental flaw in the almost flawless theory of Newton.

Einstein's general theory of relativity pictures the sun as warping space–time so as to create "hills" and "valleys," and the planet Mercury as periodically climbing the space–time "hills" in the neighbourhood of the sun. The other planets do not orbit close enough to the sun to experience any appreciable "climbing." Taking this privileged motion of Mercury into consideration, Einstein predicted in 1915 that the perihelion of Mercury should advance some 43 seconds of arc per century beyond that predicted by Newton's theory. Einstein thus brought theoretical prediction again into agreement with observation.

Space–Time and the Law of Motion

In the general relativistic view of the universe, space-time is not flat, but consists of a multitude of "hills" and "valleys." The heights of the "hills" and the depths of the "valleys" are determined by the presence of physical matter. The universal law that governs the motion of all bodies is: *All bodies follow the shortest four-dimensional space-time path between any two events.*

Thus, in this approach to the universe, the task of the physicist is somewhat reminiscent of that of Christopher Columbus. The physicist, too, sets out on a long voyage; this time, however, to chart a space–time map of the entire universe. For to know the space–time structure of the universe would be to know the behavior of all objects therein—from the photon to the giant galaxies. It is for this reason that gravity can now be viewed as a geometric property of space–time.

In Einstein's own words:

> The great attraction of the theory is its logical consistency. If any deduction from it should prove untenable, it must be given up. A modification of it seems impossible without destruction of the whole.
>
> No one must think that Newton's great creation can be overthrown in any real sense by this or by any other theory. His clear and wide ideas will forever retain their significance as the foundation on which our modern conceptions of physics have been built.[2]

[2] A. Einstein, *Out of My Later Years* (New York: Philosophical Library, 1950).

QUESTIONS

12–1. Suppose a projectile, instead of light, is fired horizontally in the spaceship accelerating upward. What kind of a path would it follow? What kind of a path would a projectile follow when it is fired horizontally on the surface of the earth?

Answer: parabolic path

12–2. Comment on Galileo's contributions to Einstein's work.

12–3. How does the "weightlessness" experienced by astronauts in orbiting spaceships illustrate the principle of equivalence?

12–4. Suppose a spaceship attains coasting motion in a straight line somewhere in outer space. Then the pilot would experience weightlessness. How is the weightlessness experienced in outer space different from that experienced in an orbiting spaceship?

12–5. Suppose a female astronaut weighs 100 pounds on a scale fixed in an elevator. If the elevator is allowed to fall freely, i.e., the rope holding the elevator is cut, what would the scale register for the astronaut's weight?

Answer: zero

12–6. Outline the arguments leading to Einstein's deduction of space curvature.

12–7. Argue that the observer at the center of Einstein's relativistic merry-go-round is in the same inertial frame as the ground-based observer.

12–8. Suppose a spaceship is directed to a "dark" region in outer space, where no stars are visible. The spaceship is engineered to maintain a steady course at a constant speed. If a slowing rate of the pilot's heartbeat is suddenly observed, what possible cause would you attribute to the slowed rate? Explain why the "darkness" of the region would not necessarily imply the absence of physical matter.

12–9. Explain why the space–time of the special theory of relativity is flat.

The General Theory of Relativity————————————————

SUGGESTIONS FOR FURTHER READING

ATKINS, K. R. *Physics,* second edition. New York: John Wiley & Sons, Inc., 1970. (Chapter 25.)

Einstein, A. "On the Generalized Theory," *Scientific American* (April 1950).

FORD, K. W. *Basic Physics.* Waltham, Mass.: Blaisdell, Publishing Co., 1968. (Chapter 22.)

GAMOW, G. "Gravity," *Scientific American* (March 1961).

13
Newton's Theory of Gravitation

No physical theory will ever overthrow the creation of Sir Isaac Newton (1642–1721). If a passion for all-pervasive general principles dominated Einstein's thinking as a way to comprehend the universe, Newton, some two centuries earlier, had also lived such passion, creating a theory of gravitation that has been described as the "greatest generalization achieved by the human mind."

Newton's Laws of Motion

Galileo seemed to take a particular delight in demolishing Aristotelian physics. One of Aristotle's claims was that uniform horizontal motion on the flat surface of the earth would require an external influence to sustain it. Galileo, on the other hand, proffered a diametrically opposite position. He argued that an object in horizontal motion on the surface of the earth would continue to move at a constant speed only in the absence of external influences.

Consider a perfectly flat, smooth inclined plane. Set a perfectly smooth, hard, bronze ball in uniform motion. If the plane is tilted so that it has an upward slope, the moving ball would climb the slope to come eventually to rest. To keep it at a constant velocity, one would need to continue to push it upward. On a downward slope, a ball would gain speed.

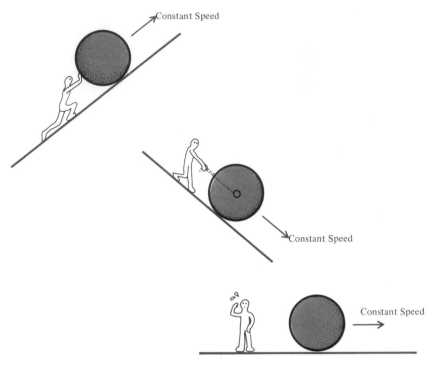

FIGURE 13-1.   Galileo's rebuttal to Aristotle.

If the ball is to be kept from gaining speed, one would have to retard the ball continuously. Thus, an external agent would be required if the ball is to be maintained in uniform motion on an inclined plane. Suppose now that the plane is adjusted so as to be perfectly horizontal. The ball would then move at a constant speed, requiring no external agent to sustain its uniform motion.

Galileo further argued that an object initially at rest would remain in the same state as long as there are no external agents to disturb them. He believed that these conclusions were a natural abstraction from observation, and summarized them in the form of a physical law. Referred to as the law of inertia, it may be stated as: *An object, initially at rest or in uniform motion, would continue in the same state of rest or of uniform motion, in the absence of all external disturbances.*

The art of abstracting physical principles from experimental observation was Galileo's special gift. He did not, however, seek to generalize them beyond the domain of immediate observation. Newton, on the other hand, recognized in Galileo's law of inertia a general principle of nature.

He perceived the principle of inertia to govern not just the motions of terrestrial objects, but motion in the entire universe as well. He thus presented it as his first law of motion: *Every body continues in its state of rest, or of uniform motion in a straight line, unless it is compelled to change that state by forces impressed on it.* Notice that the law of inertia does not explain why an object, when undisturbed, coasts in a straight line forever. It only claims that such is the way it is.

But few objects, if any, remain in a state of rest or in a state of uniform motion. They encounter influences that affect and change their states of motion. Newton, upon recognizing the existence of diverse external influences, brought them under a single concept—the concept of force. Force is defined as any physical influence that causes a change in the state of an object's motion.

In order to establish a quantitative relationship between force and the change of motion, Newton introduced the concept of mass. Mass is a measure of inertia. Inertia, in turn, is a measure of resistance to change. For example, if an object suffers a smaller change in its motion than another object under the influence of a similar force, the former object is said to have a greater mass.

Newton observed that given an object of definite mass, different forces cause different rates of change in the motion of the object. As the quantity of motion is defined as the product of mass and velocity, change in motion would imply a change in velocity, because mass is assumed to remain constant. The change in motion is such that a greater force would cause the velocity of an object to change at a faster rate. The rate of change of velocity thus becomes a concept that is essential in a description of motion under the application of a force. The rate of change of velocity is called acceleration. It is in terms of mass and acceleration that Newton stated his second law of motion:

$$\text{force} = \text{mass} \times \text{acceleration}$$

The second law of motion says, in essence, that an object under the application of a force would accelerate in the direction of the applied force with the magnitude of the acceleration completely determined by the applied force and the mass of the object.

But would anything happen to the supplier of force? Newton formulated his third law of motion to provide an answer: *To every action there is always an equal and opposite reaction.* A kicked stone, for example, would, in turn, exert an equal and opposite force on the foot.

Newton's Theory of Gravitation_____

Kepler's Empirical Laws of Planetary Motion

It was a full century before Newton's birth that Copernicus put forth the idea of a heliocentric universe. The idea was a revolutionary departure from the generally accepted Aristotelian cosmology, and thus generated controversies on all intellectual fronts. The issue, however, was essentially a scientific one, and the resolution was to be achieved ultimately in careful observation and imaginative analysis.

It was Tycho Brahe (1546–1601), a Danish astronomer of considerable wealth, who took up the challenge of observation. From his own island near Copenhagen, he observed the positions and movements of the planets, painstakingly recording what he saw. If his data seemed to reveal no immediate patterns in the diverse motions of the planets, he was, nevertheless, sustained by the faith that his work was still a right step in the direction of resolving the mysteries of planetary motion.

The man destined to decipher the mysteries was Tycho's assistant, Johann Kepler (1571–1630). Kepler sought a geometrical description of planetary motions, proceeding in this pursuit along the path of trial and error. By 1609 he succeeded in discovering that:

Law 1. *Planets describe elliptical orbits about the sun, with the sun located at one of the foci of the ellipse.*

Law 2. *The radius vector sweeps out equal areas in equal intervals of time.*

The radius vector is a line that joins the sun and the planet. In order for

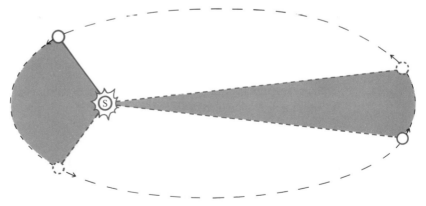

FIGURE 13–2.   Kepler's first two laws of planetary motion.

Physics: The Fabric of Reality

it to sweep out equal areas in equal times, a planet would speed up upon approaching near the sun and slow down as it gets farther away from the sun. This law is a description of the observation that the planets do not orbit the sun at a uniform speed and is referred to as the law of equal areas.

Kepler also noticed in Tycho's data that the farther away a planet is from the sun, the longer it takes for it to complete a revolution. He published this observation about planetary motion in 1619:

Law 3. *The square of the period of revolution of a planet is proportional to the cube of its semimajor axis.*

These three laws constructed order out of chaos. They expressed a system of uniformities and regularities in planetary motion. They, however, were empirical laws, thus lacking explanatory powers. Only theories are capable of such powers. A deeper understanding of planetary motion, therefore, required a fundamental theory.

Universal Gravitation

Newton's first law of motion, the law of inertia, suggested that since they do not move at uniform speeds in straight lines, the planets must be under

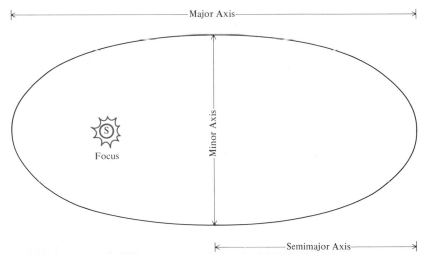

FIGURE 13–3. Definition of the major axis of an ellipse.

Newton's Theory of Gravitation_____

the influence of a force. Beyond its existence, the law of inertia revealed no clues as to the direction or the magnitude of the force.

In arriving at the direction of the force governing planetary motion, Newton was guided by Kepler's law of equal areas. Consider, to start with, a simulation, at least in appearance, of the orbital motion of a planet. A circular motion results, when a stone, attached to the end of a string, is whirled about a fixed point. The only force acting on the stone is one supplied by the holder of the string, and the force is directed inwardly along the string. Newton, pondering the nature of planetary motion, ventured a guess that the force acting on the planet might also be directed toward the sun. He then discovered that such an assumption led to Kepler's law of equal areas, thus enhancing the correctness of the guess.

A force that is directed toward a fixed center is called a central force. A central force always leads to Kepler's law of equal areas. Since central forces play an important role in many areas of physics, the law of equal areas is given another, more general name. It is called the law of the conservation of angular momentum. Kepler's second law of motion may then be stated as: *The planets conserve angular momentum.* As for the magnitude of the force, Newton found a clue in Kepler's first law. He was able to show that in order for the planets to describe elliptical orbits about the sun, the force had to become weaker as the distance from the sun increased. More specifically, the force had to vary inversely as the

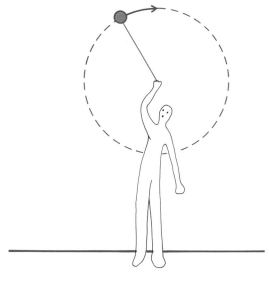

FIGURE 13–4.   A stone in circular motion.

square of the distance from the sun. In seeking to deduce Kepler's third law, Newton discovered that the force exerted on a planet also had to be proportional to the mass of the planet.

Newton thus deduced the nature of the force that governs the planets. The force had to be directed to the sun, inversely proportional to the square of the distance from the sun, and directly proportional to the mass of the planet. In symbols,

$$F \propto \frac{m_{\text{planet}}}{r^2}$$

where r denotes the distance from the sun to the planet.

The preceding law was sufficient to describe planetary motion. But Newton went one step further, and asked if the attractive force provided by the sun depended at all on the mass of the sun itself, as it did on the mass of the planet. That it did was deduced by applying his third law of motion: *to every action there is an equal and opposite reaction*. The result was that the force between the sun and a planet must be proportional to the product of the masses of the sun and the planet, and inversely proportional to the square of the distance between them.

$$F \propto \frac{M_s M_p}{r^2}$$

Now, why does an object fall? If one assumes it falls because the earth pulls it toward the center of the earth, then the moon must also fall toward the earth. So argued Newton. He knew that an object released near the surface of the earth falls a distance of 16 feet at the end of the first second. The moon's distance from the center of the earth was known to be about 240,000 miles. Newton predicted that if the moon is held by the earth via an inverse square force, the moon would fall about 0.05 inch during the first second. The result agreed beautifully with the observed fall of the moon. Newton was thus led to the conclusion that the earth was also a seat of an attractive inverse square force and that the same force causing an apple to fall also held the moon in its orbit about the earth.

Newton generalized that the earth's attractive pull on the moon was fundamentally the same as the sun's attractive pull on the planet. Going one step further, he envisioned the whole universe to be held by a similar force. He thus proposed his law of universal gravitation: *Every body attracts every other body with a gravitational force, which varies directly*

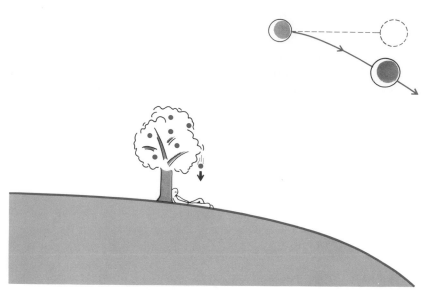

FIGURE 13–5.   If an apple falls, the moon must also fall.

as the product of the masses and inversely as the square of the distance between them. Symbolically,

$$F = \frac{Gm_1m_2}{r^2}$$

where m_1 and m_2 are the masses of the bodies interacting, r is the distance of separation, and G is the universal gravitational constant.

Measurement of G

The first measurement of the gravitational constant G was carried out in 1797 by Henry Cavendish. The idea was to hang by a fine fiber made of quartz a rigid rod with two balls attached to its ends. When two large lead balls were brought close to the smaller balls, the gravitational attraction caused the quartz fiber to twist. The amount of the twist yielded the gravitational force between the balls. From the measured values of the force, the masses of the balls, and the distance of separation, the gravitational constant was calculated. The currently accepted value is $G = 6.670 \times 10^{-13}$ mks units.

FIGURE 13–6. Newton's law of universal gravitation: $F = Gm_1m_2/r^2$.

Knowledge of the gravitational constant leads to a determination of the mass of the earth. When a body of mass m falls because of the earth's gravitational pull, the gravitational force is given by

$$F_{gravity} = \frac{GM_{earth}\, m}{R^2}$$

where R is the radius of the earth. As it falls, it accelerates, for, according to Newton's second law of motion, all bodies must accelerate under the influence of a force. Symbolically, if the acceleration of gravity is denoted by g,

$$F_{gravity} = mg$$

Equating the two expressions for $F_{gravity}$ leads to

$$g = \frac{GM_{earth}}{R^2}$$

Thus, if the radius of the earth is known, a measurement of the gravitational acceleration on the surface of the earth would yield the mass of the earth, which is $M_{earth} = 6 \times 10^{24}$ kilograms.

Newton was born into a world that was very much in intellectual bewilderment. It was a world still in shock, a world where controversies were slow to fade concerning the Copernican idea proposed some two centuries earlier. The world after Newton, however, was a triumphant one, exuding confidence in the power of pure thought. Described as the "greatest genius that ever lived" by the generations to come, Newton was held in esteem that approached adulation. No scientific theory, it may indeed be said, has had intellectual repercussions comparable to Newton's monumental work. Alexander Pope's exultation would be as appropriate today as when it was first uttered:

> Nature and Nature's laws lay hid in night:
> God said, *Let Newton be!* and all was Light.

Newton's Theory of Gravitation_____

QUESTIONS

13–1. Aristotle claimed that a force must be exerted on an object if its motion is to be maintained uniformly even on a horizontal surface.
   a. Under what circumstances is the claim valid?
   b. When is the claim not valid?

13–2. Can you cite an example where a claim drawn from the obvious can be shown to be false? Explain.

13–3. Is the law of inertia obvious in light of common experience? Explain.

13–4. What is the justification for extending the law of inertia to govern motion in the entire universe?

13–5. Motion along a circular path results when an object is acted upon by a constant force directed always toward the center of the circular path. Such an object moves also at a constant speed. Does the object also accelerate? Explain.

   Answer: accelerates toward the center of the circle.

13–6. Suppose a planet of mass m revolves around the sun in a circular orbit under a gravitational force.
   a. If the acceleration of the planet is given by $a = v^2/R$, where v is the speed of the planet, write Newton's second law of motion for the planet.

   $$Answer: \frac{GM_s m}{R^2} = \frac{mv^2}{R}$$

   b. If the period of revolution T is defined as the time the planet spends in making one complete revolution about the sun, find a relationship between the period and the radius of the circular orbit.

   $$Answer: v = \frac{2\pi R}{T}$$

   c. From (a) and (b), show that the square of the period of revolution is proportional to the cube of the radius.
   d. What is the proportionality constant?

   $$Answer: \frac{4\pi^2}{GM_s}$$

13–7. The weight of a person is equal to the gravitational attraction between the person and the earth, i.e., weight $= GM_Em/r^2$, where m is the mass of the person and r is the distance between the person and the center of the earth. How much would a person weigh at 4000 miles above the surface of the earth if his weight is 200 pounds on the surface of the earth. (Take the radius of the earth to be 4000 miles.)

Answer: 50 pounds

13–8. Assuming that tides are primarily a result of the moon's gravity, attempt an explanation of why there are two tides a day.

SUGGESTIONS FOR FURTHER READING

FEYNMAN, R. P. *The Character of Physical Law*. Cambridge, Mass.: The M.I.T. Press, 1967.

FORD, K. W. *Basic Physics*. Waltham, Mass.: Blaisdell Publishing Co., 1968. (Chapter 12.)

MARION, J. B. *A Universe of Physics*. New York: John Wiley & Sons, Inc., 1970. (Chapter 2.)

14

Electric Charge at Rest and the Electric Field

The physical world is overwhelmingly governed by the laws of electricity and magnetism. From the violent thunderstorms to the gentle hues of little flowers, from the blue sky to the white snow, much of our physical environment is comprehensible as electromagnetic phenomena. The electromagnetic phenomena are richly diverse, but beyond the diversity there lies a common cause, which is the presence of electric charge.

Electric Charge

Electric charge is the ultimate cause for all electromagnetic phenomena. The laws of electricity and magnetism describe the interactions between charged objects. An object is charged when its constituents are charged. The electromagnetic interaction, at the fundamental level, is thus an interaction between fundamental electric charges. Electric charge, however, does not exist independently of the fundamental constituents of matter. It has never been separated from matter itself. Electric charge, therefore, is simply regarded as an intrinsic property of a fundamental particle.

A great deal of mystery surrounds the electric charge. One of the mysteries is that electric charge comes in integral multiples of a fundamental unit of charge, *e*. All observed charges are thus expressible as *ne,* where *n* denotes an integer. The integers specifying the electric charge of an object, however, can be either positive or negative, as both positive and negative charges occur in nature.

The electron and the proton are fundamental constituents of matter. The electron carries a negative unity (-1) of the basic charge, and the proton carries a positive unity $(+1)$ of the same basic charge. Very sensitive experiments have been performed to determine any deviation of the proton's or the electron's charge from unity, but no deviation has been observed. The charge of the proton is precisely equal to the charge of the electron in magnitude. This is truly remarkable in light of the fact that the proton is so fundamentally different from the electron, with a mass some 2000 times as great as the electron mass.

Another mystery of the electric charge is that it is conserved. The total charge of the universe remains the same. No process, however violent, will change the total electric charge. A simple example of the law of charge conservation is the creation of an electron-positron pair from pure energy. Pair creation is essentially a violent process. One starts with a photon as a carrier of energy. The initial total electric charge is zero. As the photon is converted into a pair of electron and positron, the sum of the created charges is still zero, because the electron and the positron carry a negative and a positive unity of the basic charge, respectively.

$$\gamma \rightarrow e^- + e^+$$
$$\text{charge:} \quad 0 = (-1) + (+1)$$

There is an interesting twist to the conservation of electric charge. The theory of relativity dictates that if charge is to be conserved, the electron and the positron must be created at the same position in space. If they were created with a spatial separation between them, they would not constitute a simultaneous event to all inertial observers. Nonsimultaneity would imply that an electron, for example, would appear before a positron, and a net charge of negative unity would, briefly, be observed. Thus a violation of charge conservation would result. Consistency with relativity, then, requires that charge must be conserved locally. A pair of electron and positron, for example, cannot be created except at a common point in space. This argument of relativity can be applied to any conservation law. Relativity thus reveals that if a particular quantity is to be conserved, it must be conserved locally.

Coulomb's Law of Electrostatic Force

The gravitational force is always attractive. The electrical force, on the other hand, can be either attractive or repulsive. Repulsion exists between like charges, and attraction exists between unlike charges.

The first precision experiment to establish the exact nature of the electrostatic force was performed by the French physicist Coulomb in 1785. He began by constructing a sensitive torsion balance. A torsion balance is an instrument for precision measurements of force. A force causes a fiber, often made of quartz, to twist, and the angle of the twist yields a measure of the force. Coulomb's torsion balance consisted of a light insulating rod suspended by a thin wire, with a charged body placed at one end of the rod. When another charged body was brought near the original one, the electrical force between them would cause the wire to twist. Precision measurements of the angle of the twist of the wire were achieved by means of a mirror attached to the other end of the insulating rod.

Coulomb first observed the dependence of the force upon the distance of separation between two charged bodies. With the charges fixed, he needed only to vary the distance. He was thus led to the conclusion that

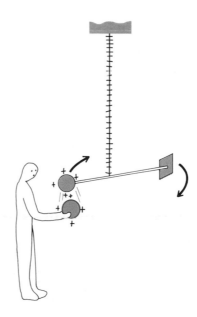

FIGURE 14–1. A schematic of Coulomb's torsion balance.

Electric Charge at Rest and the Electric Field_____

133

electric force F is, like gravitation, inversely proportional to the distance of separation r.

$$F \propto \frac{1}{r^2}$$

The problem of establishing how the electric force varied with the amounts of charge involved was a more difficult one, especially in light of the fact that he had no way of measuring the charge of anything. However, he knew how to divide the macroscopic charge. A conducting sphere, when charged, attains a uniform distribution of charge on the surface. If a similar uncharged sphere is brought into contact with a charged sphere, the original charge divides equally between the spheres, and when they are separated, each sphere has half the original charge. By repeatedly dividing the original charge, Coulomb was able to vary the charge by relative amounts. Holding the distance fixed between the charged bodies and varying the relative amounts of charge, he then was able to establish that the force was proportional to the product of the charges on each body.

$$F \propto Q_1 \times Q_2$$

Coulomb could not conceive of any other fundamental factor that would affect the force between two charged bodies. He therefore concluded that the electric force between two charged bodies is proportional to the product of the charges and inversely proportional to the square of the distance of separation between them. Symbolically, this law of Coulomb's can be expressed as:

$$F = \frac{Q_1 Q_2}{r^2} \qquad \text{(in cgs units)}$$

Note how Coulomb's law is structurally similar to Newton's law of gravitation.

Concept of the Electric Field

Coulomb's law describes quantitatively how a charged object at the point of observation is influenced by another charged object at a finite distance away. The electrical force, viewed in this way, has what one may call an unlocal character and is clearly a type of action-at-a-distance. A charge at one point knows somehow what other charges are doing elsewhere, and is thus influenced by them in a precise way. The reader may wonder if there exists between charged objects an invisible messenger which eludes a direct observation. Historically, the conceptual difficulties associated with action-at-a-distance were brushed aside by asserting that the law of electrostatic force, as formulated by Coulomb, is sufficient to describe what is observed.

There is, however, an alternative to the notion of action-at-a-distance. This is the field view of the electrical force. A field is defined to be any physical quantity that assumes a definite value at every point in space. A familiar example of a field is temperature. The value of the temperature can be obtained at any point in space by placing a thermometer there. An important aspect of a field is that its value must be independent of the measuring device, so that a field may be regarded as a property of space itself.

In the field approach to electric force, space is viewed as being imbued with electrical influences, and a charged object is viewed as interacting with the influence locally at the point of its location. The electrical influences in space are referred to as an electric field. The electric field is a special kind of field called a vector field. A vector field is specified by a definite direction as well as a magnitude at every point in space. A probe called a test charge is used to establish an electric field. A test charge is, by definition, a positive charge, whose magnitude is small so as not to disturb the field itself. If there is an electric field present at a point of observation, a test charge placed there would experience an electric force, and accelerate in the direction of the field. The magnitude of the acceleration would, then, yield a measure of the field strength, and the direction of the acceleration would yield the direction of the electric field, More specifically, the field strength is given by the electric force divided by the magnitude of the test charge.

$$\text{electric field} = \frac{\text{electric force on test charge}}{\text{magnitude of test charge}}$$

or

$$E = \frac{F}{q}$$

Such a definition does indeed insure that the electric field, like any other kind of field, will be independent of the probe.

A field always has a source. A source of the electric field is electric charge. The electric field produced by a charged object is proportional to the magnitude of the charge and varies inversely as the square of the distance away from the source.

$$\text{electric field} = \frac{\text{magnitude of source charge}}{(\text{distance})^2}$$

or

$$E = \frac{Q_s}{r^2}$$

The influence of a source charge wanes, so to speak, as one gets farther away from it.

As a way of illustrating the utility of the field concept and its equivalence to Coulomb's law, let us calculate the electrical force that a charge, say Q, would experience at a distance R away from the source charge. The force is given by the product of the charge Q and the electric field evaluated at $r = R$.

$$F = QE \text{ (at } r = R)$$

or

$$F = \frac{QQ_s}{R^2}$$

The last expression is exactly the same as Coulomb's law.

The electric field is graphically represented by directed lines. The lines of field are drawn in such a way that a test charge placed at any point would move in the direction of the field. The field lines representing the electric field of a positive point charge, for example, are directed radially outward.

Electric Potential

Suppose we arrange a pair of parallel metal plates such that one is directly above the other and the distance of separation between them is small compared to the dimensions of the plates themselves. Suppose further that the plates are connected to the positive and negative terminals of a battery. A test charge would then go from the positive plate to the neg-

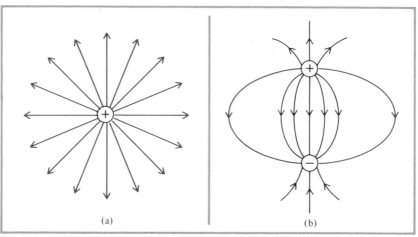

FIGURE 14–2. (a) The electric field lines of a positive charge; (b) the electric field of a charge configuration consisting of a positive and a negative charge.

ative plate along a straight line, and the force it experiences would be independent of its position. Such a field is called a uniform electric field and is represented by a series of equidistant parallel lines.

Let us now examine the motion of a test charge in a uniform field. If it is released at the positively charged plate, the test charge would accelerate along a field line toward the negatively charged plate. The test charge would thus gain kinetic energy. The situation is similar to the motion of a falling object. In the case of an object under the influence of gravity, an object, held at a certain height above the ground, is said to be at a higher gravitational potential. Upon release the object falls to a position of lower potential, gaining in kinetic energy. The gain in kinetic energy in each case represents the conversion of potential energy. In general, kinetic energy is gained whenever an object goes from a state of higher potential to a state of lower potential, and the gain in kinetic energy equals the loss in potential energy. From the motion of the test charge, then, we would deduce that the test charge is in a state of higher potential energy when it is held at the positively charged plate. The positively charged plate is, therefore, said to be at a higher electric potential. Similarly, the negatively charged plate is at a lower electric potential. There is thus an electric potential difference between the charged plates, and the kinetic energy gained by the test charge is proportional to the potential difference. More precisely, the gain in kinetic energy is equal to the product of the

Electric Charge at Rest and the Electric Field_____

magnitude of the charge and the potential difference. The electric potential difference is measured in volts in the MKS system of units.

A convenient and widely used unit of energy in atomic physics is the electron volt. An electron volt of energy is equivalent to the kinetic energy that an electron gains when it falls through a potential difference of one volt.

Lightning

The earth and the atmosphere constitute a huge battery. On a clear day the atmosphere lies between two charge concentrations of opposite sign. The ground contains negative charges, and the upper atmosphere at an altitude of 50 kilometers contains a high concentration of positive

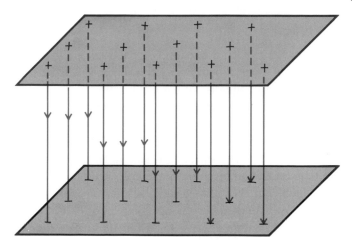

FIGURE 14–3. A uniform electric field in the region between two oppositely charged metal plates.

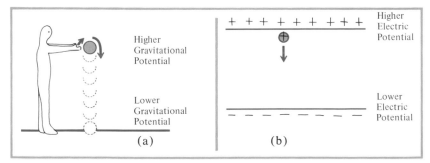

FIGURE 14–4. (a) Gravitational potential; (b) electric potential.

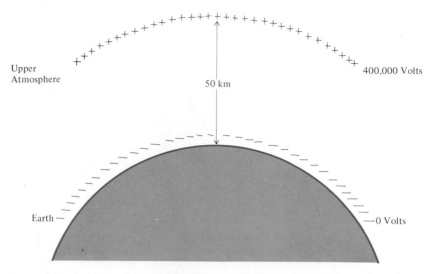

FIGURE 14–5.   Electric potential difference in the atmosphere on a clear day.

charges. The potential difference between the charged upper atmosphere and the earth's surface is about 400,000 volts. Such a potential difference implies that any positive ions (positively charged molecules and atoms) in the atmosphere would be pushed down to the ground, discharging the earth. It is estimated on theoretical grounds that the earth, if left alone, would completely discharge in about half an hour, and a disastrous electrical imbalance would result. Such a disaster, however, is prevented by a naturally occurring recharging process called thunderstorms and lightning. There are about 300 thunderstorms a day throughout the surface of the earth.

A thunderstorm begins with a rising parcel of humid air. As it rises, the air expands and cools. Upon cooling, the water vapor condenses. But the condensing vapor liberates heat, and the parcel of air becomes warmer than the surrounding air. As it continues to rise, the parcel entrains cooler air from the environment, and ice crystals are formed. The parcels form a thunderstorm cell, and the rising thunderstorm cell eventually reaches an updraft velocity of about 60 m.p.h.

When the ice crystals and water drops become too heavy, a downdraft begins. (Those who have been in a thunderstorm may recall a cool breeze, on an otherwise hot, humid day, just prior to a thunderstorm.) As the drops fall, the lower sides of the drops become positively charged and the upper sides become negatively charged, although the total charge of the drops still remains neutral. The negative ions in the atmosphere are

Electric Charge at Rest and the Electric Field⸻⸻⸻⸻⸻⸻

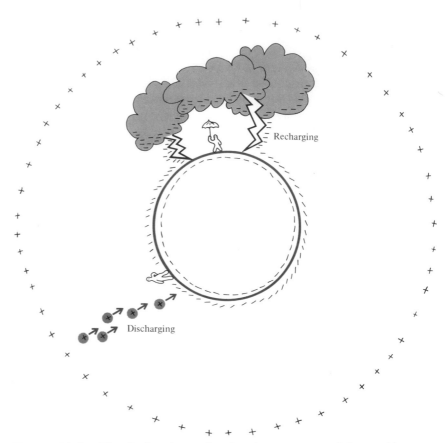

FIGURE 14–6. The discharging and recharging processes of the earth's surface.

then attracted to the lower ends of the drops, and the water drops become negatively charged. The positive ions, on the other hand, are repelled by the lower ends of the drops, and carried up by an updraft of the surrounding air. A mechanism of this type accounts—at least in part—for the fact that the bottom of the thunderclouds is usually negatively charged.

The potential difference between the negatively charged bottom of a thundercloud and the ground can sometimes rise to as high as 100 million volts. Such a potential difference implies an intense electric field in the atmosphere. But such an intense field literally drives electrons from the tops of trees and buildings to the ground, leaving the exposed objects pos-

FIGURE 14–7. A thunderstorm front. (Courtesy of S. W. Schultz.)

FIGURE 14–8. Lightning. (Courtesy of S. W. Schultz.)

Electric Charge at Rest and the Electric Field

itively charged. The buzzing sound that the mountain climber hears during a thunderstorm, for example, is due to a slow discharge from the pointed leaves and rocks. Also, the strong electric field produced by a thundercloud tears electrons from air molecules, thus ionizing the air. The ionized air serves as a conducting path for the electrons of the cloud. As the electrons dump into the earth, lightning flashes occur. The initial conducting path corresponds to a faint lightning stroke. A faint stroke is followed by a greater dumping, which begins near the ground and continues up, constituting what we call a return lightning stroke. The return stroke is what we see as a bright lightning bolt.

QUESTIONS

14–1.   A magnet consists of a south pole and a north pole—even at the sub-atomic level. Dirac has pointed out that although a monopole (an isolated south pole or north pole) has never been observed, the present state of physical theories provides no reason why a monopole could not exist. In fact, if a monopole did exist in nature, its charge g would be related to electric charge e as follows: $eg = n (\frac{1}{2} \hbar c)$, where $\hbar$ and c are constants, and n is an integer. What implications, then, would the existence of a monopole hold for an electric charge? [See K. W. Ford, *Scientific American* (December 1963).]

14–2.   Discuss the local character of mass–energy conservation, using the example of electron–positron pair creation.

14–3.   The law of charge conservation is an absolute law in the sense that no physical process has ever been observed to violate it. Which of the following production processes are forbidden by charge conservation?

a. $\gamma^{\circ} + p^{+} \rightarrow p^{+} + \pi^{-} + \pi^{\circ}$
b. $\gamma^{\circ} + p^{+} \rightarrow n^{\circ} + \pi^{+} + \pi^{\circ}$
c. $\pi^{-} + p^{+} \rightarrow n^{\circ} + \pi^{\circ}$
d. $p^{+} + p^{+} \rightarrow p^{+} + p^{+} + p^{+} + p^{-}$

Here γ, p, π, and n denotes a photon, a proton, a pion, and a neutron, respectively. p^{-} denotes an antiproton.

14–4. If the distance of separation between two charged particles is increased by a factor of two, how much would the electric force between them increase or decrease?

Answer: decrease fourfold.

14–5. Give a qualitative argument to explain the observation that charges always reside on the surface of a metal sphere.

14–6. a. If the electric field strength of a point source charge is E_0 at a distance r_0, what is the field strength at a point three times as far?
b. What is the field strength at an infinite distance away?

$$Answer: \frac{E_0}{9}$$

14–7. What is the electric field at the midpoint of a line joining two identical source charges?

14–8. Explain the principle behind a lightning rod.

14–9. a. Find an expression for the ratio of the Coulomb electrostatic force to the gravitational force between two charged particles.

$$Answer: \frac{q_1 q_2}{GM_1 M_2}$$

b. The numerical value of the ratio for two charged pions, which have a mass intermediate between the electron and the proton, is about 10^{40}. What can you say about the role of gravitation in the atomic domain?

14–10. It is observed that a charged ebonite rod can pick up bits of paper that are neutral. Assuming that atoms can be polarized by an electric field, i.e., the opposite charges making up a neutral atom can be displaced within the atom, can you explain the observation?

14–11. Explain why a water drop falling in the atmosphere may become polarized.

Electric Charge at Rest and the Electric Field⎯⎯⎯⎯⎯⎯⎯⎯⎯⎯⎯⎯⎯⎯

SUGGESTIONS FOR FURTHER READING

ATKINS, K. R. *Physics*, second edition. New York: John Wiley & Sons, Inc., 1970. (Chapter 14: "The Concept of a Field.")

FEYNMAN, R. P., R. B. LEIGHTON, and M. SANDS. *The Feynman Lectures on Physics,* Vol. II. Reading, Mass.: Addison-Wesley Publishing Co., Inc., 1963. (Chapter 9: "Electricity in the Atmosphere.")

UMAN, A. M. *Lightning*. New York: McGraw-Hill Book Company, 1969.

Physics: The Fabric of Reality

15
Moving Charges and the Magnetic Field

Electric charge at rest produces only an electric field. A moving charge, on the other hand, produces a magnetic field as well as an electric field. According to the theory of relativity, then, the magnetic and electric fields cannot exist as independent concepts. Rather, they are different aspects of a single physical entity, the electromagnetic field.

Relativity and Electric Current

Suppose a copper wire is connected to the positive and negative terminals of a battery, so that an electric current would flow in it. A current flows, because the "conduction electrons" of copper are driven by the potential difference of the battery. Electric current is simply the rate of the flow of conduction electrons. In a conductor the atom may be viewed as consisting of two parts: the outermost electron, which is so loosely bound to the atom that it may essentially be regarded as unbound or free, and the rest of the atom, which forms a tightly bound, positively charged "core." These free electrons are called conduction electrons, since their flow constitutes an electric current. The atomic cores, on the other hand, remain essentially stationary, and contribute nothing to the current.

Although the atomic cores are positively charged and the drifting electrons are negatively charged, a conducting wire, as a whole, is neutral in electric charge. The neutrality of a current-carrying conductor may be visualized by dividing up the wire into a large number of tiny volume elements, where each volume element contains equal numbers of atomic cores and conduction electrons at any time. As many electrons would flow in as flow out of each volume element. One would then expect that a test charge placed in the neighborhood of a current-carrying wire would experience no electric force from the electric current.

Observation indicates that if a test charge is set in motion relative to the wire, it does experience a force. (Let us assume that the conduction electrons are drifting to the right in the conductor frame of reference.) This force is clearly a new type of force, as it cannot be an electrical

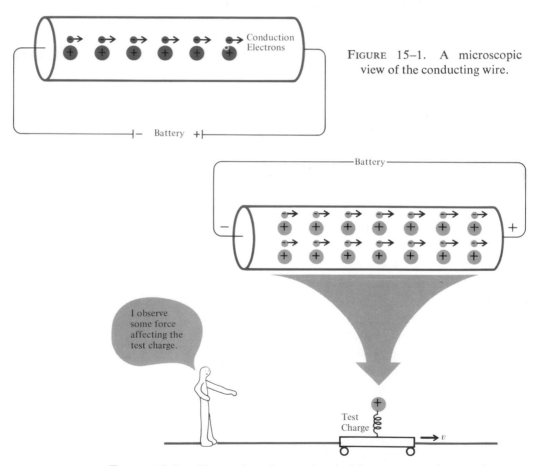

FIGURE 15–1. A microscopic view of the conducting wire.

FIGURE 15–2. Observation of a nonelectrical force on a moving test charge.

force. It is, however, intimately related to an electric current, rising and falling with the current.

As a way to discover the source of this new type of force, let us view the situation from another frame of reference. The simplest frame is one that is attached to the moving test charge. In such a frame of reference, the test charge is at rest and the current-carrying conductor is in motion. To be specific, the conductor is assumed to move to the left relative to the new reference frame, so that both the atomic cores and the conduction electrons also move to the left. But the atomic cores would be moving to the left at a higher speed than the conduction electrons. The result is that the distance between the atomic cores contracts more than the distance between the conduction electrons. Each volume element would then contain more positively charged cores than the negatively charged electrons, and would thus appear positively charged. The entire conductor, as a sum of the volume elements, would also appear positively charged in this new frame of reference. The test charge would, therefore, experience an electrical force when viewed from its rest frame of reference.

That a test charge in motion relative to a current-carrying conductor experiences a force is an experimental fact. The force, however, is clearly nonelectrical in the conductor frame of reference. This nonelectrical force, which has its origin in an electric current, is called a magnetic force. In a frame of reference attached to the test charge, on the other

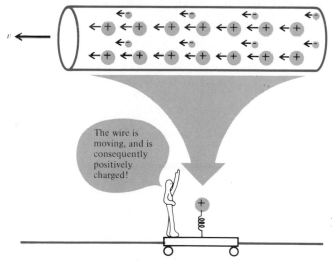

The wire is moving, and is consequently positively charged!

FIGURE 15-3. Observation of a force in the test charge frame of reference.

Moving Charges and the Magnetic Field_____

147

hand, the test charge experiences an electrical force, which arises from the relativistic effect that the conductor appears electrically charged. Thus the observed force experienced by a test charge moving in the neighborhood of a current-carrying wire could be viewed as a magnetic force in one frame of reference or as an electric force in another. In still a third frame of reference, the observed force would have to be viewed as a mixture of both electric and magnetic forces. The theory of relativity thus reveals an intimate link between the electric and magnetic fields: they are simply two different aspects of a single physical entity, the electromagnetic field.

The Fields of a Moving Charged Particle

The field produced by a charge at rest is entirely electric. If the source is positive, its field lines emanate radially from the source. If the charge is set in motion, the structure of the electric field lines changes. The change in the electric field structure of a moving charge is similar to the length contraction of a moving object. The field lines in the direction of motion are shorter than those at right angles to the motion. This implies that the strength of the electric field is stronger at the sides than ahead or behind the moving charge.

The fact that the electric field structure of a moving charge is different

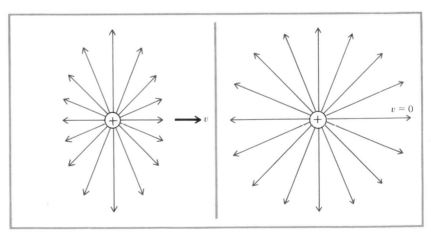

FIGURE 15–4.   The electric field structure of a moving charge.

Physics: The Fabric of Reality

from that of a stationary charge suggests that the electric field alone cannot yield a complete description of a moving charge. The field structure of a moving charge must rather be described by a mixture of both electric and magnetic fields. Put another way, a moving charge produces a magnetic field as well as an electric field.

As for the origin of the magnetic field, it is not unreasonable to guess that it owes its existence to the velocity of the charge as well as its electric field. As it turns out, the magnetic field, B, of a moving charge is simply proportional to the product of the velocity, v, and the electric field, E, of the charge. The exact mathematical expression is

$$B = \left(\frac{v}{c} \right) E \quad \text{(in cgs units)}$$

where c is the speed of light. The magnitude of the magnetic field is thus much smaller that the electric field at ordinary velocities of the charge.

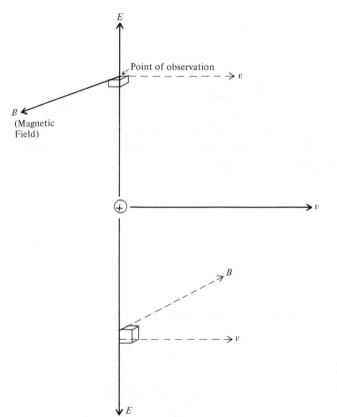

FIGURE 15–5. The magnetic field of a moving positive charge at two different points of observation.

Moving Charges and the Magnetic Field

Since a magnetic field is a vector quantity, its direction must also be specified. It turns out that the direction of a magnetic field is always perpendicular to the direction of the electric field as well as to the direction of the velocity of the moving charge. To be specific, let us examine a magnetic field at a point in space which is directly above a moving charge. The electric field at the same point is directed vertically upward. If the charge is moving horizontally, the magnetic field must be directed either out of or into the plane of the page, as the plane defined by the velocity and the electric field lines is identical to the plane of the page. The magnetic field in this example is directed out of the plane of the page. The result is that if they are viewed by an observer ahead of a moving charge, the magnetic field lines of a positive charge form circular rings about the line of motion, in a counterclockwise direction.

Magnetic Force

When a charged particle enters a magnetic field, it experiences a magnetic force. The force depends on four factors: the strength of the magnetic field, the velocity of the charged particle, the magnitude of the charge, and the angle between the direction of the magnetic field and the direction of the velocity of the particle. The force is zero when the angle is zero, and it reaches a maximum when the angle is 90 degrees. In the latter case, the magnetic force may be expressed as

$$F = \frac{q\,v\,B}{c}$$

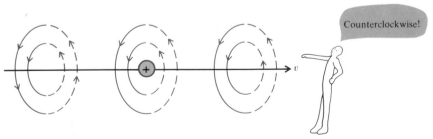

FIGURE 15–6.   The magnetic field lines of a moving charge form concentric circles.

Physics: The Fabric of Reality

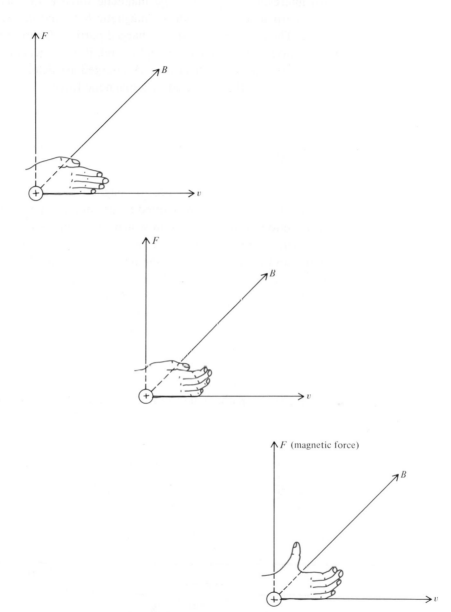

FIGURE 15–7. Magnetic force acting on a positive charge moving at right angles to the magnetic field. As the fingers wind from the velocity line to the magnetic field line, the erect thumb points in the direction of the magnetic force.

Moving Charges and the Magnetic Field

An interesting property of the magnetic force is that its direction is always perpendicular to both the magnetic field and the velocity of the particle. The force on a positively charged particle, for example, points in the direction of the thumb of the right hand, if the fingers are curved from the velocity line to the field line. A charged particle, cutting across a magnetic field, is thus deflected by a magnetic force.

Uniform Magnetic Field

A bar magnet is a practical source of magnetic fields. The magnetic field lines of a bar magnet may be plotted by use of a compass. The lines flow, by convention, in the direction in which the compass needle points. The lines emerge from the north pole of a bar magnet, and enter the south pole to join the original lines at the north pole again. The fact that magnetic field lines form closed paths is an important feature of magnetism.

Suppose the north pole of a bar magnet is placed facing the south pole of a similar bar magnet. The magnetic field lines would then go from the north pole of one magnet to the south pole of the other. The resulting lines in the region between the pole faces are straight and equally spaced, representing a uniform magnetic field.

A uniform magnetic field can be used to deflect a particle so as to move along a circular path. A uniform magnetic field exerts on a particle en-

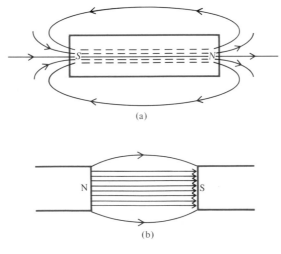

(a)

(b)

FIGURE 15–8. (a) Magnetic field lines of a bar magnet; (b) a uniform magnetic field in the region between the opposite poles.

tering the field at right angles a force that is constant in magnitude and directed perpendicular to the direction of the particle's motion. Such a magnetic force is analogous to the force that a string exerts on a stone, when the stone is tied to the string and whirled around about a fixed point. In both cases a circular motion results.

The Earth's Magnetic Field and the Radiation Belts

Most magnetic fields that occur in nature are not uniform. An example of a nonuniform magnetic field is the magnetic field of the earth. The structure of the earth's magnetic field is similar to that of a bar magnet. The field is the strongest at the poles, and grows weaker with a decrease in the latitude.

As a nonuniform field, the earth's magnetic field may serve as a "magnetic bottle" for moving charged particles. When charged cosmic ray particles enter the vicinity of the earth, they encounter the earth's magnetic field, thus experiencing a magnetic force. The force is such that the charged cosmic ray particles are driven in a spiraling motion along the lines of the earth's magnetic field. At the poles, where the lines converge, the spiraling motion of the charged particles is reversed, as the diameter of the spiral, which varies inversely with the intensity of the

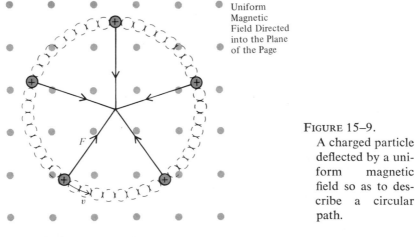

Uniform Magnetic Field Directed into the Plane of the Page

FIGURE 15–9.
A charged particle deflected by a uniform magnetic field so as to describe a circular path.

Moving Charges and the Magnetic Field

153

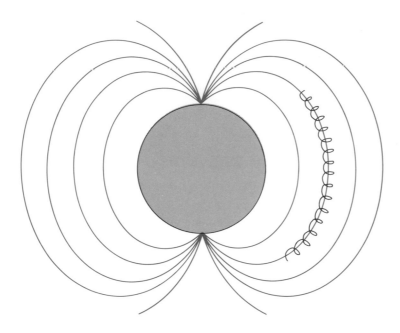

FIGURE 15–10. Spiraling motion of a charged particle along a magnetic field line of the earth.

field, ceases to shrink further. Charged particles can thus be trapped by the earth's magnetic field.

Experiments to develop a profile of the trapped cosmic ray particles were performed by J. A. Van Allen in the 1950's. Detectors were placed in artificial satellites that were launched to orbit the earth. As the readings on the detectors were broadcast back to the earth, the message was deciphered to mean that the earth is ringed by two regions of high-energy radiation, extending thousands of miles into space. The inner and outer regions are each located about 2,000 miles and 10,000 miles above the surface of the earth. They contain predominantly protons and electrons, which spiral back and forth from one pole to the other in a little over a second. These regions of radiation are now referred to as the Van Allen radiation belts.

The Aurora Borealis

The Van Allen radiation belts serve as a sort of leaky bucket for the cosmic rays that constantly bombard the earth. As the belts are refilled

Physics: The Fabric of Reality

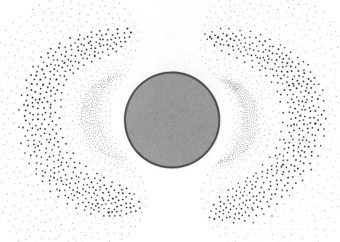

FIGURE 15–11.   The Van Allen radiation belts.

by the constant influx of cosmic rays, a leakage drains them away into the atmosphere. The leakage takes place predominantly at the magnetic poles. The most visible effects of the spilled-over cosmic rays, therefore, occur in the regions near the poles.

As these cosmic ray particles enter the atmosphere, they collide with the atmospheric atoms such as oxygen and nitrogen. Such collisions transfer energy to the atoms, thus exciting them. The excited atoms, however, quickly return to their normal states by emitting appropriate photons, some of which appear as visible light. The faint airglow in the night skies near the magnetic poles is a manifestation of the normal leakage of charged particles from the radiation belts.

Upon occasion these cosmic rays spilling into the atmosphere display spectacles of great beauty and grandeur in the night sky. The sky looks as if covered with waving curtains of various hues. Such a phenomenon, referred to as the northern lights, occurs when the atmosphere is bombarded by abnormally large influxes of cosmic ray particles. These extra quantities of charged particles often originate in the sun and get hurled toward the earth by a solar flare. As they enter the atmosphere, they excite an abnormally large number of atmospheric atoms. These excited atoms, in turn, radiate photons of different frequencies. Thus result the hazy red, white, and green hues of the aurora borealis.

Moving Charges and the Magnetic Field_____

QUESTIONS

15–1. Consider two parallel conducting wires with equal electric currents flowing in the same direction.

    a. By going to the reference frame attached to the drifting electrons in one of the wires, deduce whether the force between the conductors would be attractive or repulsive.

    b. What is the force called in the conductor frame of reference?

15–2. Establish the magnetic field of a moving particle carrying a negative charge.

15–3. A charged particle entering a uniform magnetic field at right angles to the field can be deflected so as to describe a circular path. What kind of a path would a charged particle describe if it enters a uniform magnetic field at an oblique angle? Explain. (*Hint:* Consider the extreme case where the particle enters the field parallel to it. In that event the particle would suffer no deflection.)

Answer: spiral

15–4. Suppose a uniform magnetic field is established by means of electromagnets. What would be the effect on a charged particle describing a circular path in the field region if the strength of the magnetic field is increased? Explain.

Answer: circular path becomes smaller

15–5. Consider two positively charged particles moving at a nonrelativistic velocity v parallel to each other.
  a. Show that the ratio of the electric force to the magnetic force between them is v^2/c^2.
  b. What can you say about the relative importance of magnetic force in the presence of electric force?
  c. When is magnetic force of dominant importance?

Answer: in absence of electric force

15–6. Antimatter, such as positrons, must be created in order to be observed, as it suffers annihilation upon encounter with matter. Devise a way of storing positrons.

15–7. People near the poles are subjected to higher intensities of cosmic rays than people at the equator. Explain.

15–8. Would there be auroral activities in the southern hemisphere also? Explain.

SUGGESTIONS FOR FURTHER READING

ATKINS, K. R. *Physics,* second edition. New York: John Wiley & Sons, Inc., 1970. (Chapter 16: "Moving Charges.")

Moving Charges and the Magnetic Field_____

FEYNMAN, R. P., R. B. LEIGHTON, and M. SANDS. *The Feynman Lectures on Physics,* Vol. II Reading, Mass.: Addison-Wesley Publishing Co., Inc., 1963. (Section 13-6: "The Relativity of Magnetic and Electric Fields.")

VAN ALLEN, J. A. "Radiation Belts Around the Earth," *Scientific American* (March 1959).

16

Accelerated Charges and Electromagnetic Waves

There exist in nature electric and magnetic fields that change with time. These dynamic fields cannot be produced by charges that are at rest, although certain dynamic fields can be produced by charges in uniform motion. The radiation of dynamic fields, however, can only be achieved by charges that are in nonuniform motion.

Induced Electric and Magnetic Fields

An essential feature of the dynamic fields is that they themselves can serve as the sources of other dynamic fields. It is an experimental fact that a *changing* magnetic field induces a changing electric field. Such a field must therefore be regarded as a source of electric fields. The symmetry of nature then suggests that a changing electric field might likewise induce a changing magnetic field, thus serving as a source of magnetic fields, and such, indeed, is the case.

Another feature of the dynamic fields is that a process of mutual induction can be achieved if the fields themselves propagate at a certain critical velocity. That the critical velocity must be the speed of light can be argued as follows: Suppose we start out with a changing electric field which propagates in space at a velocity, v, assumed to be less than the

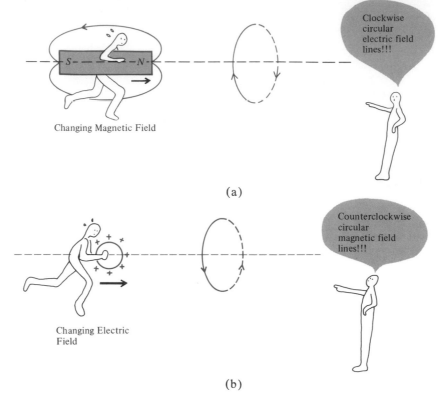

Changing Magnetic Field

Clockwise circular electric field lines!!!

(a)

Changing Electric Field

Counterclockwise circular magnetic field lines!!!

(b)

FIGURE 16–1. (a) Changing magnetic field inducing electric field; (b) changing electric field inducing magnetic field.

speed of light, c. The magnetic field induced by the changing electric field would then be less than the inducing electric field, as

$$B \text{ (induced)} = \left(\frac{v}{c}\right) E \text{ (inducing)}$$

The new magnetic field, when it in turn induces a new electric field, cannot, however, produce an electric field greater in magnitude than itself. Rather, the induced electric field would be less than the inducing magnetic field. The inducing field would thus be inducing a successively weaker field, and the process of mutual induction would eventually come to a halt. On the other hand, if the fields propagate at the speed of light, the induced field would be equal in magnitude to the inducing field. The speed of light thus serves as the critical velocity for the mutual reinforcement of traveling electromagnetic fields.

Physics: The Fabric of Reality

A third feature of dynamic fields is that an induced field is always perpendicular to the inducing field. If the inducing electric field is along the y axis, then the induced magnetic field would lie along the x or z axis. A question of fundamental significance is: How are the directions of the dynamic fields related to the direction of propagation of these fields? Maxwell's electromagnetic theory predicts that the fields must be transverse to the direction of propagation.

The transverse nature of the electric and magnetic fields propagating at the speed of light may be made more plausible by considering a charge moving at a speed close to the speed of light. The electric and magnetic fields produced by such a charge could be viewed as traveling at the same speed as the charge itself. As fields traveling at near the speed of light, they would then have a structure similar to that of a radiated electromagnetic field, which does propagate at the speed of light as a result of mutual reinforcement.

The field structure of a moving charge can be obtained by taking the relativistic length contraction into consideration. The field lines of a stationary charge can be likened to the support beams of a spherical dome. The electric field lines of a moving charge would appear as length-contracted support beams of a moving dome. A dome moving at a high speed would, of course, attain the shape of a pancake, with the longer support beams all lying in a plane perpendicular to the direction of motion. The electric field of a charge moving nearly at the speed of light would then be essentially transverse.

The direction of a magnetic field, on the other hand, is always perpendicular to the plane defined by the electric field line and the line of motion of the charge producing it. If the charge is moving along the x axis and the electric field lies along the y axis, then the magnetic field would lie

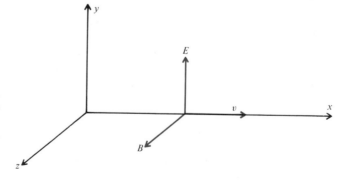

FIGURE 16–2. Electric and magnetic fields are perpendicular to each other, and both are, in turn, perpendicular to the direction of propagation of the fields.

Accelerated Charges and Electromagnetic Waves_____

along the z axis. Thus, the magnetic field of a high-speed charge is, also, essentially a transverse field.

In short, then, a charged particle traveling nearly at the speed of light produces electric and magnetic fields, which are essentially transverse. Equivalently, electric and magnetic fields traveling at such speeds must also be transverse. The transverse nature of the radiated electromagnetic fields, then, comes as no surprise as they must, by necessity of existence, propagate at exactly the speed of light.

Fields Produced by an Accelerated Charge

A charge at rest produces only a static electric field. A charge in uniform motion produces dynamic fields, but it cannot radiate dynamic fields. The radiation of dynamic electromagnetic fields can only be achieved by accelerated charges.

Let us consider an electron accelerated toward a positive metal terminal. As a charged particle in motion, the electron would set up an electric and a magnetic field. As the magnetic field is proportional to the velocity of the particle, the magnetic field would attain a maximum value just before the electron reaches the metal target. A sudden stopping of the electron by the metal target, however, would cause the magnetic field to collapse to zero. The collapsing magnetic field would then induce a new electric field, which, in turn, would induce a new magnetic field. The sudden stopping of an electron would thus radiate an electromagnetic field, and the field would carry away the kinetic energy of the electron.

Suppose the electromagnetic field is hurled out into empty space. The

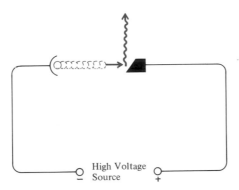

FIGURE 16–3. A moving electron, upon being stopped by the metal target, radiates an electromagnetic field.

High Voltage Source

principle of conservation of energy dictates that such a field cannot be allowed to die out, for otherwise the energy carried by the field would disappear in space with nothing to absorb it. Therefore, the process of mutual reinforcement between the electric and the magnetic fields must be sustained out in space. But such a requirement implies that an electromagnetic field radiated by an accelerated (or decelerated) charge must propagate at the speed of light.

Electromagnetic Waves

Consider now a particular type of acceleration, namely, a sinusoidal acceleration. By a sinusoidal acceleration, one means that the magnitude of the acceleration varies with time as a sine curve. Two common examples of sinusoidal acceleration are the periodic swinging motion of a pendulum and the periodic motion of an object attached to a spring.

An electric charge executing a sinusoidal acceleration radiates an electromagnetic field that varies as a sine curve with respect to time. This is so, because the magnitude of a radiated electric field is proportional to the acceleration of the charge producing it. The resulting electric field rises in magnitude from zero to a maximum value $+E_0$ and then returns to zero. Reversing its direction, it then continues on from zero to a negative maximum $-E_0$, finally returning to the original value of zero.

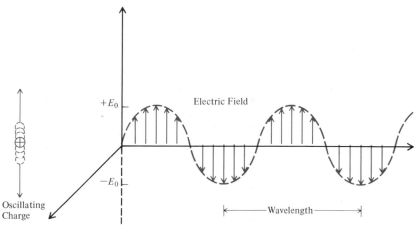

FIGURE 16–4.   Electric field produced by a sinusoidally oscillating charge.

Accelerated Charges and Electromagnetic Waves _____

An electric field completes a cycle, when starting at a particular value, it returns to the same value. The length of time required to complete a cycle is called the period of vibration. The number of vibrations per second is called the frequency of vibration. The frequency, f, and the period, T, are related by a simple inverse relationship, $f = 1/T$. The distance that the electric field traverses during one period of vibration is called the wavelength, and is commonly denoted by the Greek letter λ. The velocity of propagation, c, is then given by

$$c = \frac{\lambda}{T} \qquad \text{(definition of speed)}$$

or

$$c = \lambda f$$

An electric charge executing a sinusoidal oscillation also produces a time-varying magnetic field. Once produced, these electric and magnetic fields reinforce each other and thus sustain their propagation in empty space. The process of mutual reinforcement takes place in phase. As a changing electric field induces a changing magnetic field, the magnitude of the magnetic field rises with the magnitude of the electric field, attaining the maximum value exactly at the same time. An electromagnetic field sustained by a process of mutual reinforcement is also referred to as an electromagnetic wave.

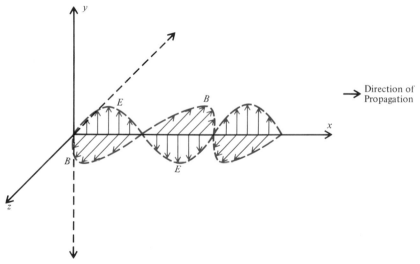

FIGURE 16–5.   Electromagnetic wave.

An electromagnetic wave must propagate at the speed of light, for otherwise the process of mutual reinforcement between the electric and magnetic fields would not be sustained. Its speed then suggests that an electromagnetic wave must be a transverse wave. As a transverse wave, the electric and magnetic fields must vibrate in directions perpendicular to the direction of propagation.

Energy, Momentum, and Mass of Electromagnetic Radiation

Suppose electromagnetic radiation falls on a charged particle at rest. The charged particle would immediately respond to the vibrating electric field of the radiation, and thus oscillate. A particle in oscillatory motion, however, has kinetic energy. Since the particle is initially in a state of zero kinetic energy, it must be surmised that energy is transferred from the electromagnetic radiation to the charged particle.

The energy of electromagnetic radiation is proportional to the product of the electric and magnetic fields. The exact expression for electromagnetic energy crossing unit area per second is

$$\text{Electromagnetic energy} = c\left(\frac{EB}{4\pi}\right)$$

where E and B denote the electric and magnetic fields. A more energetic electromagnetic radiation thus has a stronger vibrating electric field than a less energetic radiation.

Electromagnetic radiation also imparts linear momentum to a charged particle. When a charged particle oscillates in response to a vibrating electric field, it also experiences a magnetic force. This is so, because a charged particle experiences a magnetic force whenever it moves across a magnetic field. The effect of the magnetic force is to cause a particle to move along the direction of the propagation of the radiation. The particle is thus pushed forward by electromagnetic radiation, and such motion can be viewed as resulting from the transfer of momentum from the radiation to the particle. Electromagnetic momentum is also proportional to the product of the electric and magnetic fields, or the mathematical expression for electromagnetic momentum crossing unit area per second is

$$\text{electromagnetic momentum} = \frac{EB}{4\pi}$$

Notice now how marvelously simple is the relationship between the electromagnetic energy and momentum. They are linked by the speed of light.

$$\frac{\text{electromagnetic energy}}{\text{electromagnetic momentum}} = \text{speed of light}$$

Energy and momentum, in the theory of relativity, are different aspects of a single physical concept. For an ordinary particle, the energy–momentum relationship contains the rest mass of the particle.

$$\text{energy} = \sqrt{(c \times \text{momentum})^2 + (\text{rest energy})^2}$$

In the special case of zero rest mass, the relationship reduces to a very simple one.

$$\frac{\text{energy}}{\text{momentum}} = \text{speed of light}$$

But this is exactly the relationship that we obtained for the energy and momentum of electromagnetic radiation. We are thus led to the very interesting conclusion that *the rest mass of electromagnetic radiation is zero.*

QUESTIONS

16–1. Suppose a wire loop is connected to a galvanometer (a device to measure current). The galvanometer needle deflects when a bar magnet is thrust into the loop. Explain the origin of the current in the loop.

16–2. If a charge is moving at 1 per cent the speed of light, what is the ratio between the electric and magnetic fields produced by the charge?

Answer: 100:1

16–3. If a charged object is tied to a string and whirled around, would it, in principle, radiate electromagnetic fields?

16–4. If an electromagnetic wave is observed to travel at a smaller speed in a medium than in a vacuum, what can you say about its wavelength in a medium as compared to that in a vacuum? (The frequency remains the same.)

Answer: shorter

16–5. The mutual reinforcement between the electric and magnetic fields of an electromagnetic wave is maintained as long as the wave propagates at a certain critical speed.
  a. Would this critical speed of propagation be different to different observers?
  b. Relate your answer to the postulates of the special theory of relativity.

16–6. A physical entity having a zero rest mass must always exist in a state of motion, traveling at the speed of light. Why could such an entity not exist, on the basis of relativity, at a speed less than the speed of light? (*Hint*: In the relativistic equation $E = pc$, write $E = mc^2$ and $p = mv$, where m is now an effective moving mass.)

16–7. a. Explain the mechanism by which electromagnetic radiation transfers energy to a charged particle.
  b. Can electromagnetic radiation also deliver energy to a neutral particle? Explain.

16–8. What is meant by the description that electromagnetic radiation is a transverse wave?

16–9. Explain why if electromagnetic radiation did not propagate at the speed of light in vacuum, it would not conserve energy.

16–10. Does the electromagnetic wave require a medium for propagation?

SUGGESTIONS FOR FURTHER READING

FEYNMAN, R. P., R. B. LEIGHTON, and M. SANDS. *The Feynman Lectures on Physics,* Vol. II. Reading, Mass.: Addison-Wesley Publishing Co., Inc., 1963. (Section 20–3, "Scientific Imagination.)
FORD, K. W. *Basic Physics.* Waltham, Mass.: Blaisdell Publishing Co., 1968. (Chapter 18.)

Accelerated Charges and Electromagnetic Waves⎯⎯⎯⎯⎯⎯⎯⎯⎯⎯

17

The Electromagnetic Nature of Light

The most dramatic prediction of Maxwell's electromagnetic theory was the existence of electromagnetic waves. These waves, which are radiated by accelerated charges, owed their existence to a certain critical speed of propagation. This critical speed turned out to be identical to the speed of light. To Maxwell, the identity of speed between light and the electromagnetic wave was more than a coincidence; in fact, the identity of speed implied the very identity of the physical entities themselves. He was thus led to the brilliant conclusion, published in the 1860's, that light and the electromagnetic wave are one and the same thing. Light was thus to be understood within the domain of electromagnetism.

The Scattering of Light

The most fundamental aspect of the interaction of light with matter is that light, as a vibrating electromagnetic field, exerts an electromagnetic force on atomic electrons, thus causing them to perform tiny vibrations. Most of what we see around us as light-related phenomena can be understood on the basis of the forced vibrations of the atomic electrons.

The interactions of light with matter are most simply described in terms of an atomic oscillator. The oscillator model of the atom views an atom as though an electron were bound to the atomic core by a "spring." An electron so held would execute oscillatory motion as a part of its natural state. The frequency of natural oscillation is called the resonant frequency of oscillation. All atoms then possess a resonant frequency.

When light falls on matter, the atomic electrons respond to the vibrating electric field as well as to the vibrating magnetic field of the incident light. But the magnetic force exerted on an electron is much weaker than the electric force, and therefore may be ignored. In such an approximation, light may simply be treated as a vibrating electric field.

A simple type of interaction is the scattering of light. The vibrating electric field of incident light sets atomic electrons in oscillatory motion. The electrons in forced oscillation serve, in turn, as a source of electromagnetic waves. In terms of energy, light delivers energy to the atom, thus exciting it, and the atom reradiates it in the form of light. Such a process of absorption and re-emission of radiation by a charged particle is referred to as scattering. In most simple situations, incident light causes the electron to oscillate at the same frequency as itself. The oscillating electron, then, reradiates electromagnetic waves whose frequency is the same as that of incident radiation, yielding a scattered radiation identical to the incident light.

Atoms, however, are not efficient scatterers of light. They ordinarily scatter an extremely small portion of incident light. Scattering, nevertheless, can be a very important phenomenon if there is a large number of atoms causing a multiple scattering or if the intensity of incident light is large. As scatterers of light, atoms are more efficient in scattering the light of one frequency than that of another.

The Blue Sky

Light from the sun gets scattered by the atmospheric atoms, and enters our eyes in the form of scattered light. The different components of visible light suffer varying degrees of scattering. The degree of scattering that visible light undergoes is related to the frequency of incident light as well as to the resonant frequency of the scattering atom. The frequencies of visible light are much lower than the resonant frequencies of the atmospheric atoms. This fact leads to the result that the intensities of scattered light for different frequencies are essentially independent of the resonant frequencies of the atoms. Insofar as the scattering of visible light is con-

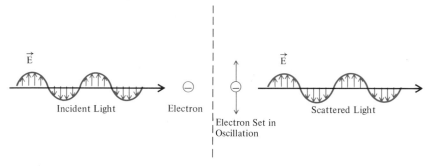

FIGURE 17–1.   Scattering of light: interaction of a vibrating electric field, E, with an atomic electron.

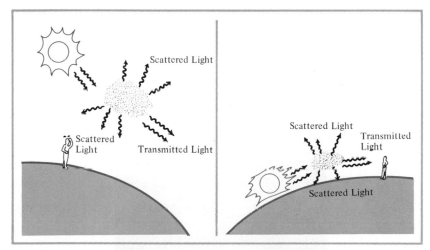

FIGURE 17–2.   Scattering of sunlight by the atmosphere (a) at noon, (b) at dusk.

cerned, then, the different kinds of atoms in the atmosphere need not be distinguished. The intensity of scattered light is simply proportional to the fourth power of its frequency.

$$\text{intensity scattered} \propto (\text{frequency})^4$$

To be specific, let us compare the scattering of the red and blue components of sunlight. The frequencies of blue and red light are about 7×10^{14} and 4×10^{14} Hz or vibrations per second, respectively. The ratio of the intensities of the scattered light, then, comes out to be:

The Electromagnetic Nature of Light

$$\frac{\text{intensity of blue scattered}}{\text{intensity of red scattered}} = \frac{f^4_{blue}}{f^4_{red}} = 10$$

The interpretation is simple and straightforward. The atmosphere scatters blue light about ten times as much as red light, yielding a blue color for the sky.

The intensity of scattered light, however, is a very small portion of the intensity of incident light, as individual atoms are not very efficient scatterers of light. The thickness of the atmosphere is not great enough so that the intensity of the sunlight is not sufficiently reduced. The sun itself is thus too bright for direct viewing, and the portion of the sky near the sun's position at high noon is white rather than blue.

FIGURE 17–3. A sunset on the southern coast of Korea. (Courtesy of E. Kim.)

The Sunset

The red color of the sunset is also a result of scattering. The sunlight from the setting sun traverses a great deal more of the atmosphere than that from the sun at noon. It thus suffers more scattering with a corresponding reduction in the intensity of the unscattered light. In fact, the intensity of the transmitted light is sufficiently reduced so that it is possible to view the sun directly as it sets over the horizon. But the atmospheric atoms scatter blue light some ten times as efficiently as red light. The transmitted light thus consists predominantly of the lower-frequency components of the spectrum, yielding a reddish-orange color for the setting sun.

The colors of the sunset, however, often blaze the whole western sky. They, too, result from the scattering of light by the atmosphere. The light initially scattered to illuminate the western sky may be more blue than red, but as it traverses the atmosphere to reach the observer, its blue hues get scattered out, leaving only the red components in the transmitted beam.

Pollutants in the atmosphere also add to the colors of the sunset. A particle whose size is of the order of the wavelength of blue light scatters blue light better than any other colors. As the size grows, however, it scatters preferentially those colors whose wavelengths are comparable to the size of the particle. Beyond the size of the wavelength of red light, a particle scatters predominantly red light.

Polarization of Scattered Light

Consider as a source of electromagnetic waves an electron that vibrates along the y axis. If we single out an electromagnetic wave that is radiated along the x axis, we shall find that the electric fields of such a wave vibrate exclusively along the y axis. Such a wave is said to be linearly polarized with the direction of polarization along the y axis. In general, a vibrating charge yields a linear polarization along the direction of its oscillatory motion.

When polarized waves superpose at random, the result is an unpolarized wave. The electric fields then do not vibrate along one direction

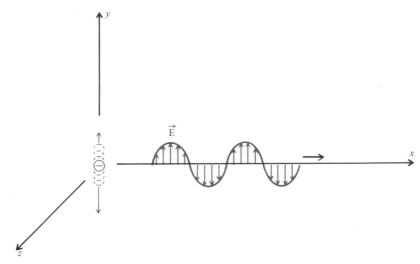

FIGURE 17–4. A linearly polarized light wave. The electric fields vibrate parallel to the y axis.

of polarization, but along all directions that are transverse to the direction of the propagation of light.

Sunlight is unpolarized, because the oscillatory motions of the electrons in the sun have random orientations. It can, however, be polarized upon scattering by atoms. Scattered light achieves complete polarization, if it is viewed at 90 degrees from the direction of the unpolarized incident light. An unpolarized incident beam traveling along the x axis can be represented by the electric fields that vibrate along the y and z axes. An electron, at the origin of a coordinate system, would then undergo oscillatory motion either along the y axis or along the z axis. An observer directly below the electron—anywhere along the negative y axis—would receive no radiation from the electron oscillating along the y axis, because an oscillating electron cannot emit radiation to propagate in the same direction as the direction of its own oscillatory motion. An oscillator emits the greatest intensity of radiation in directions that are at right angles to the direction of its oscillatory motion. Thus the only radiation the observer receives would come from an electron oscillating along the z axis, but such radiation would be linearly polarized along the z axis. Scattered radiation observed elsewhere, however, would be partially polarized, as the radiation would now consist of electric fields which vibrate along the z axis as well as along the y axis and which are unequal in magnitude.

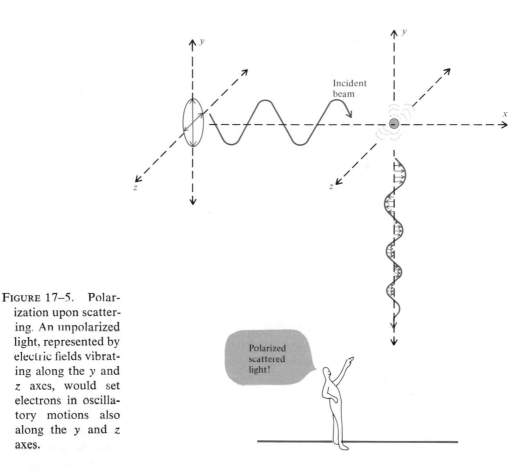

FIGURE 17–5. Polarization upon scattering. An unpolarized light, represented by electric fields vibrating along the y and z axes, would set electrons in oscillatory motions also along the y and z axes.

Reflection

Reflection results when a certain number of atomic oscillators at the reflecting surface vibrate in unison. The number of oscillators that must act in unison is determined by the wavelength of incident light, and it is proportional to the square of the wavelength. Now the intensity of light scattered by oscillators acting in unison is, in turn, proportional to the square of the number of oscillators. The intensity, therefore, varies as the fourth power of the wavelength of incident light. The intensity of light scattered by an individual atom, on the other hand, varies inversely as the fourth power of the wavelength of incident light. Both of these factors take part in giving rise to a reflected light. Consequently, the intensity of reflected light is independent of wavelength. A surface, impinged upon by white light, thus yields a reflected light that is also white.

Snowflakes expose to sunlight a surface area greater than the square of the wavelengths of sunlight. With a sufficient number of atoms thus capable of vibrating in unison, snowflakes reflect all the colors of white light with equal intensity. Consequently, snow is white.

On the other hand, if certain colorless grains are smaller in size than the wavelengths of visible light, they would not be able to expose to light a surface area great enough for the atomic oscillators therein to act in unison. The atoms would then act as individual oscillators and would scatter more blue than red light. The blue color of cigarette smoke is a familiar example of scattering of light by small grains.

Polarization upon Reflection

A polarizer is a substance that transmits only those electric fields that vibrate parallel to the axis of polarization and that absorbs all the other electric fields. Polaroid sunglasses are an example of a polarizer.

When unpolarized light is reflected off of a surface, it becomes partially polarized. The direction of the stronger electric field in the reflected light is parallel to the surface of the reflector and is therefore horizontal to an observer standing on the reflecting surface. The glare off the road is thus horizontally polarized. A reduction in the glare may then be achieved by a polarizer with a vertical axis of polarization. Polaroid sunglasses are just such a polarizer.

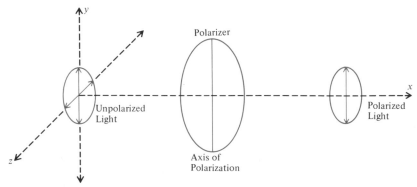

FIGURE 17–6.   Polarization of light by means of a polarizer.

Physics: The Fabric of Reality

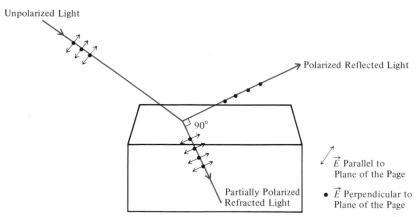

Unpolarized Light

Polarized Reflected Light

90°

$\vec{E}$ Parallel to
Plane of the Page

$\vec{E}$ Perpendicular to
Plane of the Page

Partially Polarized
Refracted Light

FIGURE 17–7.　Polarization upon reflection.

A complete polarization of reflected light occurs when it is viewed
at 90 degrees from the direction of the refracted (transmitted) beam. All
the electric fields of the reflected light then vibrate parallel to the reflecting surface.

Dispersion

The speed of light in a transparent medium is less than the speed of light
in a vacuum. The reduced speed arises from the fact that light in a
medium travels via an absorption–re-emission mechanism of the oscillator model of the atom. It takes time—something like 10^{-8} seconds—
for an atom to absorb and re-emit light. The propagation of light, therefore, suffers a corresponding retardation in transit through a medium.

The process of absorption and re-emission of light by an atom depends
upon the frequency of light. An atom absorbs more readily the frequencies that are closer to its resonant frequency. Since the resonant frequencies of atoms lie mostly in the ultraviolet region, an atom can absorb
and re-emit blue light more readily, and thus more often, than red light.
Consequently, blue light suffers a greater retardation in passing through
a medium than red light. In general, the higher the frequency of light
is, the lower is its speed in a medium.

When light enters a medium, it bends at the surface. The amount of

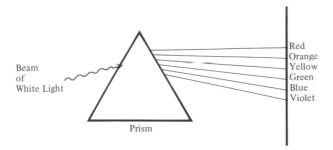

Beam
of
White Light

Prism

Red
Orange
Yellow
Green
Blue
Violet

FIGURE 17–8. Dispersion of light by a prism.

bending depends upon the speed of light in the medium. The light of a lower speed bends more. The violet color, for example, bends more than the red color. A medium, such as a prism, is then capable of dispersing white light into its component colors. A naturally occurring example of dispersion of light is the rainbow.

QUESTIONS

17–1. The violet color of the visible spectrum corresponds to a higher frequency than the blue. Why is the sky not violet?

17–2. Observe scattered light with a pair of Polaroid sunglasses. Describe and explain your observations.

17–3. Explain why a sunbather lying on her side would find that her Polaroid sunglasses are not as effective in reducing the glare as when she is sitting up.

17–4. Explain why clouds are white.

17–5. Explain why the sky above cities is a pale mixture of white and blue.

17–6. If a material absorbs every visible color except red, its color is red. Water is a weak absorber of red light, so that its intrinsic color is a pale greenish blue. Explain why a red shell would look black when it is observed at a depth of, say, 30 meters of water.

17–7. Explain why objects that do not emit light of their own are still visible.

17–8. If the wavelength of red light is given to be about twice the wavelength of blue light, what is the ratio of the number of oscillators that would be set to vibrate in unison by each light?

Answer: 4:1

17–9. Give a simple method of determining whether a given light beam is completely polarized.

17–10. List three light-related phenomena that are readily observable in your environment. Give a brief explanation of each on the basis of the electromagnetic nature of light.

17–11. Comment on the remark that physics "makes terrific demands on the imagination" in relation to electromagnetic waves.

SUGGESTIONS FOR FURTHER READING

HEWITT, P. G. *Conceptual Physics.* Boston: Little, Brown and Co., 1971. (Chapter 29.)

WEISSKOPF, V. F. "How Light Interacts with Matter," *Scientific American* (September 1968).

18

The Quantum Nature of Light

The electromagnetic theory of light is a theory of dazzling success. Viewing light as an electromagnetic wave and atomic oscillators as emitters and absorbers of light, the theory renders the universe more approachable and comprehensible. But the story of light does not end there, for the rich imagination of nature has more in store for man to explore.

Radiation from Incandescent Objects

An incandescent object, regardless of its chemical composition, emits radiation of all frequencies. The emitted radiation varies in intensity with the frequency. The intensity peak occurs at a certain intermediate frequency, and the intensity falls off toward both the higher and the lower ends of the intensity curve. The color of an incandescent object corresponds to the frequencies of the emitted radiation near the intensity peak. Now the frequency of the intensity peak is a function of temperature. The relationship is linear, so that with a rising temperature the peak frequency shifts to a higher frequency, corresponding to a change in the color of the object. An incandescent object thus glows red to blue as its temperature is raised.

The intensity curve, or spectral shape, of radiation from an incandescent object is independent of the chemical composition of the object itself.

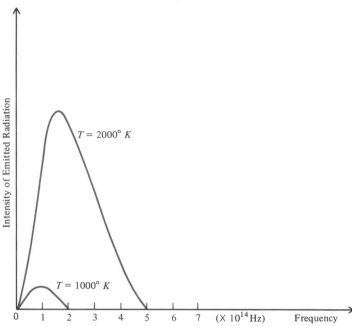

FIGURE 18–1. Temperature dependence of the intensity peak of incandescent radiation.

This fact suggests that the frequency distribution of incandescent radiation may have a common origin, namely, the atomic oscillator. A successful prediction of the spectrum would then require a model of the atomic oscillator. Maxwell's electromagnetic theory provides a model for the atomic oscillator. The spectral analysis of incandescent radiation, thus, affords another opportunity to confirm the validity of the theory. When theoretical physicists attacked the problem of incandescent radiation toward the end of the nineteenth century, they were optimistic.

Maxwell's electromagnetic theory assumes that an atomic oscillator can have any amplitude of vibration so long as its oscillatory behavior is not destroyed. The amplitude of oscillation is defined as the displacement of the oscillator from its stationary, or equilibrium, position. As the amplitude of oscillation gives the energy associated with oscillatory motion, the classical theory views an atomic oscillator to exist in any energy state, with the allowed energy states forming a continuum.

The classical model of an atomic oscillator, however, led to strange consequences. Applied to the emission of radiation from an incandescent object, the model predicted that the intensity of radiation would increase

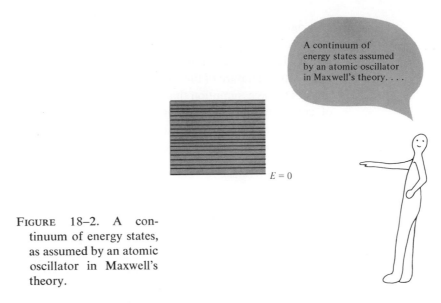

FIGURE 18–2. A continuum of energy states, as assumed by an atomic oscillator in Maxwell's theory.

with the frequency. Such a prediction was in clear disagreement with the experimental result that the intensity of radiation falls off toward the higher end of the frequencies. Furthermore, the total energy radiated by an incandescent body was predicted to be infinite, so that even in the absence of experimental results, the prediction would have been absurd. This failure of classical physics was so serious in implication that it was referred to as the "ultraviolet catastrophe."

The "ultraviolet catastrophe" came when physics was regarded as a mature discipline in need of no further conceptual development. Some physicists felt that the universe had been rendered entirely comprehensible through the monumental works of Newton and Maxwell, and that what remained ahead was no more than taking care of details which would, in no way, be of fundamental nature. A confrontation with something as devastating as the "ultraviolet catastrophe," therefore, came as a complete surprise.

The Quantum Nature of Oscillators

Max Planck (1858–1947), a German theoretical physicist, pointed a way out of the "ultraviolet" dilemma in 1900. Approaching the problem by trial and error, Planck succeeded in producing a theoretical spectrum

The Quantum Nature of Light

that agreed with the observed frequency distribution of radiation from an incandescent body. His success, however, was founded upon a revolutionary idea that was utterly foreign to classical physics. The new idea dealt with the nature of the oscillator itself and traced the error of classical physics to the assumption that an atomic oscillator may have a continuum of energy states. He discovered that in order to derive a spectrum in agreement with observation, he had to postulate that: *An atomic oscillator may exist only in discrete, quantized energy states, which are specified by the relation*

$$E_n = nhf, \ n = 0, 1, 2, \dots$$

where f is the frequency of the oscillation and h is a proportionality constant. E_n represents an allowed energy state of the oscillator, and is specified by an integer n, called a quantum number.

That an oscillator may exist only in certain discrete energy states is an idea that represents a drastic departure from common experience. Planck's idea implies for the oscillatory motion of an atomic oscillator that its amplitude of oscillation may assume only certain discrete values. To appreciate the revolutionary nature of Planck's quantum hypothesis, consider, for example, the oscillatory motion of a pendulum. If one claims that a pendulum can swing only through certain discrete angles of arc, say 5, 10, and 15 degrees, common observation would judge him to be

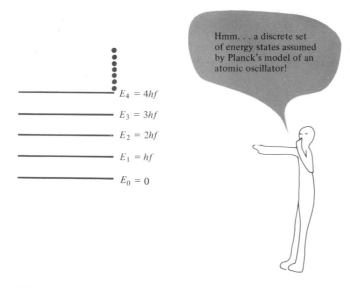

FIGURE 18–3. Planck's model of an atomic oscillator, assuming a discrete set of quantized energy states.

clearly in error. For anyone could demonstrate that a pendulum swings continuously through any angle of arc.

Strange as it was, Planck's quantum hypothesis did resolve the "ultraviolet catastrophe" of classical physics. To break with the classical concept of a continuum, however, seemed too abrupt even for Planck himself, so that he continued, albeit unsuccessfully, to direct his efforts to remove the "ultraviolet catastrophe" still within the structure of classical physics. In the meantime, however, the quantum idea continued to meet the challenges of observation where Maxwell's theory failed most dramatically. One such example was a readily demonstrable phenomenon called the photoelectric effect.

The Quantum Nature of Light

The photoelectric effect, first observed by Heinrich Hertz in 1887, refers to the release of electrons from a metal surface upon irradiation by light. The application of Maxwell's electromagnetic theory to the photoelectric effect was simple and straightforward. The vibrating electric field of incident light would set the electrons in the metallic surface in oscillatory motion, thus imparting energy to the electrons. The amount of energy so imparted would be proportional to the magnitude of the electric field. Since the electric field strength also determines the intensity of light, a brighter incident light would deliver a greater amount of energy to the electrons. Furthermore, energy delivered by an electromagnetic wave would be

FIGURE 18–4. The photoelectric phenomenon.

cumulative, so that even a dim light would release electrons from a metal surface, if it radiated the surface long enough. It was calculated that with readily available intensities of light, an exposure time of a few minutes would be required to eject electrons with kinetic energies of several electron volts.

Observation, however, was in disagreement with these predictions of the electromagnetic model of light. It was found, for example, that red light did not release electrons from a zinc surface, no matter how bright the light was and no matter how long the exposure lasted. Apparently red light simply could not deliver the energy required to release an electron. Another experimental result was that the photoelectrons carry an energy of the order of a few electron volts. Such electron energies would require a few minutes of exposure to electromagnetic waves. However, the release of the photoelectrons was observed to be immediate (a delay time of less than 10^{-9} seconds was later established). It was thus clear, to the dismay of physicists, that the electromagnetic theory of light had failed to explain the photoelectric phenomenon.

It was in this setting that Einstein, in 1905, proposed a theory of the photoelectric effect based on the quantum hypothesis of Planck. Planck's postulate states that atomic oscillators exist in quantized energy states, specified by the equation

$$E_n = nhf, n = 0, 1, 2, \ldots$$

where h is Planck's constant and f is the frequency of oscillation. Einstein argued that when an oscillator makes a transition from a higher energy state to a lower state, e.g., from E_2 to E_1, the oscillator would emit a quantum of electromagnetic radiation with an energy content of hf, which is simply the difference between E_2 and E_1. Such a quantum of electromagnetic radiation is now called a photon.

Einstein viewed a photon as localized in space, thus having a particlelike character rather than a wave nature. Its energy content was assumed to be proportional to the frequency of radiation with Planck's constant as the proportionality constant. That is,

$$\text{Energy of photon} = \text{Planck's constant} \times \text{frequency}$$

or

$$E = hf$$

He further asserted that the photon traveled at the speed of light, retaining its localized particlelike character as it moved in space.

Einstein applied his photon model of light to the photoelectric effect. He envisioned a photon colliding with an electron bound to a metal surface and delivering its entire energy to the electron. If the photon energy is sufficient to free an electron from the binding force of the surface, an electron would be released. The kinetic energy of a photoelectron would then be simply the initial photon energy less the binding energy. Symbolically,

energy of photon − binding energy = kinetic energy of photoelectron

or

$$hf - \text{B.E.} = \text{K.E.}$$

Notice that the kinetic energy of the photoelectron is directly proportional to the frequency of the incident photon. The relationship then can be tested experimentally by measuring the kinetic energies of the released electrons while varying the frequencies of the light. Such a measurement would also yield the value of Planck's constant in the equation.

The numerical value of Planck's constant was obtained from a study of the photoelectric effect by Robert A. Millikan in 1916. The measured value agreed with the value of Planck's constant obtained from Planck's analysis of radiation from an incandescent object. It was thus clear that both the incandescent radiation and the photoelectric effect, although they seemed to be unrelated phenomena superficially, are in essence quantum phenomena, with Planck's constant as a linkage.

Planck's quantum hypothesis thus received a big boost in the hands of Einstein. Einstein was proposing that not only the oscillator but light itself must be quantized. Light must now be viewed as a particle whose energy is determined by the frequency of light. A break with the electromagnetic theory of light, at last, seemed imminent as well as inevitable.

The Compton Scattering

A simple yet crucial test of the photon model of light may be provided by a scattering process. It is well known that electromagnetic radiation is scattered by charged particles. Scattering in the electromagnetic theory of light is viewed as a process in which incident radiation sets a charged particle in oscillation and the oscillating particle, in turn, reradiates

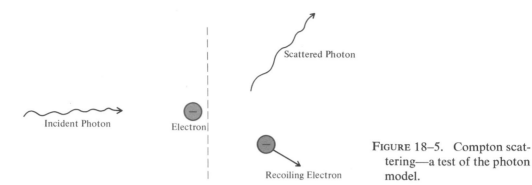

FIGURE 18–5. Compton scattering—a test of the photon model.

electromagnetic waves. In the case of a free electron as a scatterer, the frequency of the incident radiation is the same as the frequency of the scattered radiation.

The photon model of light views scattering in a different way. A particlelike photon collides with a charged particle, say an electron, and imparts a part of its initial energy to the electron. The scattered photon then goes off in some direction, but with a reduced energy. But a reduced energy implies a reduced frequency for the scattered photon. Thus, the photon model predicts that scattered light would have a smaller frequency than incident light.

In 1922 Arthur H. Compton performed a scattering experiment. He knew that the electromagnetic nature of X rays had been demonstrated in other experiments. His objective was to determine whether X rays behaved also like a photon. He bombarded a metallic foil with a beam of X rays of well-defined frequency. He then measured the frequencies of the scattered X rays in various directions. He discovered that the scattered X rays came off with reduced frequencies, as predicted by the photon model of radiation. When he applied the photon model to an analysis of the experiment, the agreement between theory and experiment was complete. Here was an unequivocal demonstration of the validity of the photon model.

The Dual Nature of Electromagnetic Radiation

Is electromagnetic radiation a wave or a particle? The electromagnetic nature of light emerged as a consequence of Maxwell's requirement that the laws of electricity and magnetism be mutually consistent. The electromagnetic wave model of light has since been demonstrated in numerous experiments and has been applied successfully to most of the natural phenomena involving light.

The photon model of light, on the other hand, emerged when the electromagnetic theory of light failed to account for certain readily demonstrable phenomena. They made sense only when light was viewed as a particle—as a localized packet of energy.

Both the electromagnetic wave and the photon obey Einstein's relativistic energy-momentum relation. Just as the energy of electromagnetic radiation is simply the product of its momentum and the speed of light, the energy of the photon is also given by a similar equation,

$$E = pc$$

where p denotes the momentum of the photon. The rest mass of the photon is, therefore, zero, and the photon, as a massless particle, always travels at the speed of the light. Both models of light are thus consistent with the special theory of relativity. The physicist's picture of the physical world therefore requires that light be viewed both as a wave and as a particle.

QUESTIONS

18–1. For a given incident photon energy, not all photoelectrons have the same kinetic energy. Why would this be so? Explain.

18–2. The numerical value of Planck's constant is

$$h = 4.15 \times 10^{-15} \text{ eV-sec}$$

If the frequencies of red and blue light are 4×10^{14} vibrations/sec, and 7×10^{14} vibrations/sec, respectively, what are their energies in electron volts?

Answer: 1.66 eV and 2.91 eV

18–3. Explain why a good emitter of radiation would also make a good absorber of radiation.

18–4. If the energy binding an electron to the surface of potassium is 2.20 electron volts, what minimum energy must a photon have in order just to free a photoelectron? Also, calculate the frequency of the photon.

Answer: 5.3×10^{14} vibrations/sec

18–5. If the incident photon, in a Compton scattering experiment, has an energy of 1 million electron volts, and the scattered photon has 0.3 million electron volts, how would you account for the energy difference?

18–6. Comment on the role of models in the physicist's mode of operation.

18–7. Discuss the role of models in your own discipline.

18–8. Suppose a beam of red light is observed to carry the same amount of energy as a beam of blue light. How would you account for this in light of the fact that a photon of blue light has a higher energy than a photon of red light?

Answer: more photons of red light

SUGGESTIONS FOR FURTHER READING

WEIDNER, R. T., and R. L. SELLS. *Elementary Modern Physics*. Boston: Allyn and Bacon, Inc., 1960. (Chapter 3.)

19

Light from Hydrogen

Throw some common salt into the flame of a Bunsen burner and the flame will take on a bright yellow color. The color of the strontium flame is red, and that of thallium is green. The colors are characteristic of the elements themselves.

Spectral Lines of Atomic Hydrogen

When an electric discharge is passed through a hydrogen container, the hydrogen atoms emit radiation. The radiation consists of several different colors, and they form a part of the hydrogen spectrum.

Suppose light from a hydrogen source is made to pass through a narrow rectangular slit. The image of the slit would then be also of rectangular shape. But such an image would provide no information concerning the color components of the spectrum. If we recall that different colors correspond to different frequencies, the dispersive property of a prism lends itself for the resolution of colors. If the light passing through a slit is made to pass through a prism, the slit image would, then, be colored, and the images of different colors would appear at different positions on the screen. If the slit is narrow enough, the images on the screen would constitute a series of lines. It is for this reason that a spectrum from an atomic source is often called a *line spectrum*. And the components of the spectrum are referred to as *spectral lines*.

A prism spectrometer resolves the hydrogen spectrum into four readily visible lines, whose colors are red, blue, blue-violet, and violet. For sunlight, on the other hand, the same spectrometer yields a continuous

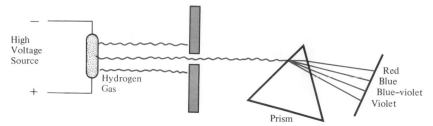

FIGURE 19–1.   Visible spectral lines of hydrogen.

spectrum, with the colors dispersed adjacent to each other with no spatial separation between them. The line spectrum of hydrogen, therefore, stands in sharp, perhaps mysterious, contrast to the more familiar continuous spectrum of sunlight.

Balmer's Empirical Formula

The wavelengths of the four visible spectral lines of hydrogen were accurately measured by A. J. Ångström, a Swedish physicist, during the latter half of the nineteenth century. When accurate experimental results exist, they tempt theoretical physicists to seek out patterns and regularities in them. As the astronomical observation of Tycho Brahe had done for Johann Kepler some three centuries earlier, Ångström's spectral lines revealed a pattern of amazing simplicity to Johann Balmer in 1885.

Balmer discovered that the frequencies of the four hydrogen spectral lines were all fractional multiples of one common factor. The fractions were 5/36 for red, 3/16 for blue, 21/100 for blue-violet, and 2/9 for violet. At first sight these fractional numbers corresponding to different lines do not appear to form a pattern, but they can be expressed in the following pattern.

$$f_{red} = K\left(\frac{1}{2^2} - \frac{1}{3^2}\right)$$

$$f_{blue} = K\left(\frac{1}{2^2} - \frac{1}{4^2}\right)$$

$$f_{b\text{-}v} = K\left(\frac{1}{2^2} - \frac{1}{5^2}\right)$$

$$f_{violet} = K\left(\frac{1}{2^2} - \frac{1}{6^2}\right)$$

Physics: The Fabric of Reality

or

$$f = K\left(\frac{1}{2^2} - \frac{1}{n^2}\right), \qquad n = 3, 4, 5, 6, \ldots$$

The frequencies represented by Balmer's formula agreed with the experimental values with striking precision.

He then calculated the frequency of a hypothetical fifth line. When he later learned of Ångström's measurement of the fifth line, his theoretical value did not agree with the experimental value with the same precision as the previous four lines. The discrepancy, however, was slight, and he thought that some external influence on the hydrogen source might be the cause. At any rate, he was intrigued by the prospect that a general formula may exist to represent the spectral lines of all atoms.

Balmer's formula, however, yields no picture for the hydrogen atom. What, for example, do the integers 2 and 3 have to do with the red light that the hydrogen atom emits? Balmer's work was a good beginning, but the answers to such questions can only be provided by a model of the atom.

Thomson's Atomic Model

Light from hydrogen, and from other heated elements, lent strong support to the idea of the electrical constitution of matter. But atoms are normally neutral in electric charge. Thus it was reasonable to view them as consisting of equal amounts of positive and negative charge. The question as to whether electric charge is associated with some fundamental constituents of matter was, however, still unresolved in the latter part of the nineteenth century.

In 1897 J. J. Thomson provided convincing evidence for the existence of an elementary particle with which negative charge was to be associated. The evidence emerged from his study of the so-called cathode rays which emanated from the negative terminal of a high-voltage source. He was able to measure a ratio of the ray's mass to its charge and found that it was about 2000 times smaller than a similar ratio for a hydrogen ion. These rays carried negative charge, whereas hydrogen ions were positively charged.

The smallness of the ratio of mass to charge for the cathode rays could be due either to a small mass of the rays or to a large charge. Thomson found it more reasonable to assume the former and deduced that the mass of the ray is less than the mass of a hydrogen ion by a factor of 2000. He called the cathode particle a corpuscle and hypothesized it to be a fundamental constituent of matter. Thomson's corpuscle is now called an electron.

His work on the electron led Thomson to propose a model for the atom. He envisioned a neutral atom as consisting of electrons embedded within a spherical distribution of positive charge. The mutual repulsion among the electrons would then cause them to be uniformly distributed throughout the sphere of positive charge. This picture was often referred to as a "raisin-cake" model of the atom.

The attractive feature of the model was that it provided a qualitative explanation for the emission of light from atoms. Thomson was able to show that in order for the vibrating electrons to produce visible light, the atoms would have to have a radius of about 10^{-8} cm, in good agreement with the size of the atom deduced from the kinetic theory of gases. However, attempts to calculate a spectrum in agreement with observation were not as fruitful. No one, for example, succeeded in deriving Balmer's formula from Thomson's model. A lack of quantitative success thus rendered the model less attractive, although that did not constitute a proof that the model was wrong. To prove it wrong required a more decisive test.

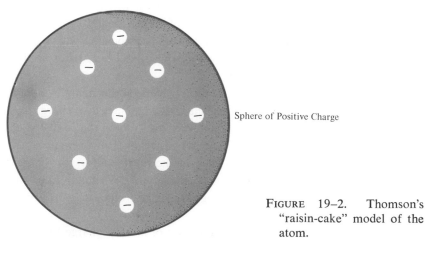

Sphere of Positive Charge

FIGURE 19–2. Thomson's "raisin-cake" model of the atom.

Rutherford's Alpha-Particle-Scattering Experiment

One way of probing an atom is by firing an atom-sized particle to collide with it. Thomson's model predicts exactly how certain probe particles would scatter off of the target atoms. Thus, a scattering experiment would provide a crucial test for the Thomson model of the atom.

E. Rutherford embarked on just such an experiment about 1910. For the target atoms he chose a thin foil of gold. He chose gold because it is amenable to extreme thinness. A thin foil assured that fewer atoms would be encountered in the path of the projectile particles, and would thus enhance the chance of single collisions between the projectile and the atom. For if a projectile scatters off of many different atoms, the information revealed about a target atom would be less direct.

For his projectile Rutherford used the so-called alpha particles, which are emitted at a high speed from certain radioactive sources. The alpha particle was known to be about 8000 times as massive as the electron and to carry a positive charge whose magnitude is twice the charge of the electron. The alpha particles, on the other hand, were very light compared to the gold atoms, their mass being only about one fiftieth the mass of the gold atom.

Thomson's atomic model predicted that an alpha particle traversing a gold foil would suffer a deflection by gold atoms and that the deflection would be caused by the positive charge distribution as well as by the embedded electrons. The positive charge distribution, however, would cause only a very slight deflection, because it is uniformly distributed over a sphere of radius 10^{-8} cm, which is some 100,000 times the size of an alpha particle. The electrons would also cause only a very small deflection, because they are some 8000 times as light as the alpha particles. The result is that the alpha particles would suffer little deflection and would scatter predominantly in the forward direction.

The experimental setup consisted of an alpha source, two diaphragms to collimate the alpha particles into a narrow parallel beam, a gold foil, and a detector of the alpha particles. The experiment was carried out by Rutherford's two able assistants, Geiger and Marsden.

The experimental result was a complete surprise. About one in every 20,000 of the incident alpha particles was completely bounced back by a gold foil only 4×10^{-5} cm thick. This was a far cry from the prediction of the Thomson model that no alpha particles would deviate more than

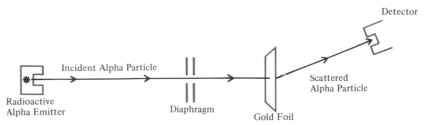

FIGURE 19–3.   A schematic of Rutherford's alpha-particle-scattering experiment.

at most one degree from the incident path. So unexpected was the result that Rutherford is reported to have commented, "It was quite the most incredible event that has ever happened to me in my life. It was almost as incredible as if you had fired a 15-inch shell at a piece of tissue paper and it came back and hit you."

Rutherford reported his findings in 1911, concluding that the assumption of a uniform distribution of positive charge in Thomson's model was totally untenable. Rather, he proposed that the positive charge is concentrated at the center of an atom, within a region of radius 10^{-12} cm. He called the positive core the nucleus of the atom, and worked out a theory of alpha particle scattering on the basis of his nuclear atom. In the nuclear model, the alpha particles collide with the massive nucleus, and therefore scatter in all directions, including the backward direction. Rutherford suceeded in deriving a formula to describe the observed alpha-scattering results and thus put the hypothesis of the atomic nucleus on a sound basis.

Rutherford's Nuclear Atom

Armed with unquestionable experimental evidence for a nuclear atom, Rutherford proceeded to explore its further implications. First of all, he argued that the concentration of positive charge in the atomic nucleus implied the existence of the electrons in the outer region of the atom. This outer spherical region was to determine the overall size of the atom. Then there was the question of the force holding the electrons in this region. He assumed that the force was the well-known electric force between opposite charges. Thus, the atom was to be governed by the laws of electricity and magnetism. But an attractive force would pull the electrons toward the center of the force. This situation necessitated postulating a circular motion for the electrons, with the result that the nuclear atom emerged as a miniature solar system.

A planetary atom would emit radiation according to Maxwell's electromagnetic theory. Recalling the emission of radiation from atoms, one might, at a first glance, be awed by the widening sphere of Rutherford's successes, but a storm gathered quickly. An orbiting electron accelerates toward the nucleus continuously, and would, therefore, radiate a continuous spectrum of light. Atoms, however, radiate discrete spectral lines. Maxwell's electromagnetic theory of light is unbending on this point. An orbiting electron must emit radiation with a frequency determined by the frequency of revolution of the electron. But as it radiates energy, the electron would have less energy, and therefore would be pulled in closer to the nucleus. As it continues to emit radiation of continuously increasing frequency, the electron would finally collapse to the nucleus. Thus, Maxwell's theory predicted for Rutherford's planetary atom a total collapse and a continuous spectrum, with both results in violent contradiction with observation.

The dilemma was not easy to resolve. There was ample experimental evidence for both Rutherford's nuclear atom and Maxwell's electro-

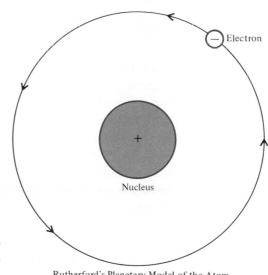

FIGURE 19–4. Rutherford's planetary model of the atom.

Rutherford's Planetary Model of the Atom

Light from Hydrogen_____

magnetic theory. Furthermore, Maxwell's theory was the crowning achievement of theoretical physics as a synthesis of the separate sciences of electromagnetism and optics. The situation was thus quite desperate.

Bohr's Quantum Atom

It was Niels Bohr's good fortune to cross the English Channel in 1911 as a brand-new Ph.D. It was perhaps even more fortuitous that he came under the influence of Rutherford. There in Manchester he quickly found himself caught up in the excitement of Rutherford's group. It was right in their midst that the nuclear atom was being shaped into perfection by each new alpha particle arriving at the detector. When he was shortly called back to Copenhagen to a lectureship in theoretical physics, he took back with him the wonder of the nuclear atom.

To Bohr, now in Copenhagen, the defect was not with the nuclear atom, but with Maxwell's electromagnetic theory. He pinpointed the defect to Maxwell's prediction of a continuous spectrum from a continuously oscillating system. Planck had already shown that an oscillator of atomic size does not oscillate in a continuous manner with just any frequency, but only with a discrete set of frequencies. When the nuclear atom was dressed with Planck's quantum clothes, Bohr discovered that the nuclear atom emerged as a viable physical system.

Bohr started out by keeping part of Maxwell's theory. He assumed that the nucleus attracts the electron with the well-known electrical force. He then borrowed Newton's laws of motion, so that the orbiting electron would be governed by them. But he rejected the prediction of Maxwell that such an orbiting electron would emit radiation. Instead, he turned to Planck's quantum hypothesis.

The quantum hypothesis implied, for the nuclear atom, a set of discrete, quantized energy states. As long as the electron remains in one of these quantized states, so argued Bohr, the atom would not emit radiation. To stress the nonradiating character of these states, he called them the stationary states. The electron in a stationary state, then, continues to orbit about the nucleus, but does not radiate. It was this assumption that represented a break with Maxwell's theory, constituting a first step in the quantum approach to the atom.

The energy of the stationary states of the hydrogen atom was calculated to be:

$$E_n = \frac{-E_0}{n^2}, \ n = 1, 2, 3, \ldots$$

E_n denotes the energy of an nth stationary state, n is called a quantum number, specifying a particular quantum state, and E_0 is a constant, having a numerical value of 13.6 electron volts. The negative sign is intended to convey the notion that a hydrogen atom is a bound system.

The orbit of the electron was also found to be quantized. Since the size of an orbit may be specified by a radius, the quantized orbits may be given by quantized radii:

$$r_n = n^2 r_0, \ n = 1, 2, 3, \ldots$$

Again, r_n denotes the radius of an nth orbit, and r_0 represents the normal size of the hydrogen atom. Bohr's calculations yielded a numerical value of

$$r_0 = 0.53 \times 10^{-8} \ \text{cm}$$

Bohr then assumed that the electron may make a transition from one stationary state to another stationary state. Such a transition would in-

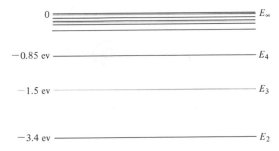

FIGURE 19–5. The energy states of the hydrogen atom, as predicted by the Bohr model.

Light from Hydrogen

volve a change in the energy content of the atom. The energy difference in a downward transition, for example, would be carried away in the form of a quantum of radiation. The observed spectral lines would then appear as manifestations of the existence of discrete energy states in the atom and as direct results of downward transitions between states.

To be specific, let us consider a transition from the $n = 3$ state to the $n = 2$ state. The energy of the quantum of radiation resulting from such a transition would be given by:

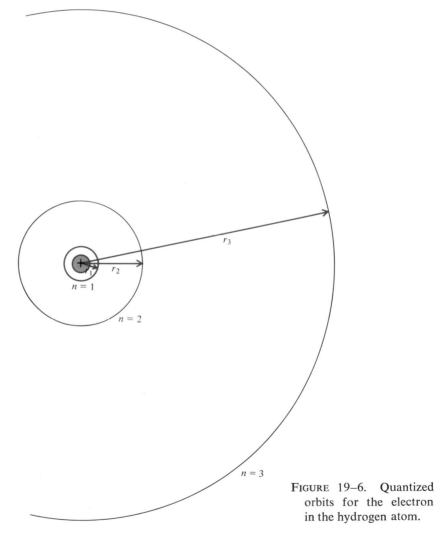

FIGURE 19–6. Quantized orbits for the electron in the hydrogen atom.

$$\text{energy of photon} = E_3 - E_2$$

But the energy of a photon is simply the product of the frequency of radiation and Planck's constant. Therefore

$$hf_{3 \rightarrow 2} = E_3 - E_2$$

or

$$f_{3 \rightarrow 2} = \frac{E_0}{h}\left(\frac{1}{2^2} - \frac{1}{3^2}\right)$$

where $f_{3 \rightarrow 2}$ denotes the frequency of the emitted photon.

Now compare Bohr's expression for the frequency of radiation emitted upon a transition from the third orbit to the second orbit with Balmer's empirical expression for the frequency of red light from hydrogen. They are identical! The previously mysterious presence of the integers in Balmer's empirical formula is thus accounted for by Bohr's quantum model of the atom.

Hydrogen atoms normally exist in the lowest energy state, called the ground state, E_1. Atoms in the ground state do not radiate, because such would follow a still further downward transition, but lower energy states do not exist. Emission of radiation, therefore, involves initially an upward transition to higher energy states from the ground state. An upward transition is called excitation, and all the states higher than the ground state constitute the so-called excited states. Excitation is achieved by supplying

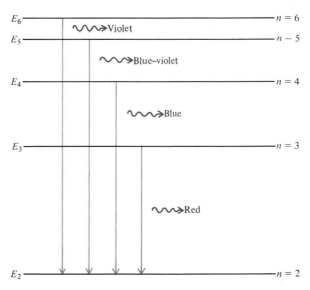

FIGURE 19–7.   The Bohr model description of the visible spectral lines of the hydrogen atom.

Light from Hydrogen

energy to the atom from an external source. One method of excitation is the use of an electric discharge, which essentially subjects atoms to a high potential difference.

If the supply of energy is sufficient, the electron of a hydrogen atom may be dissociated from the hydrogen nucleus. Such dissociation is called ionization, and the minimum energy required to separate an electron from its parent atom is called ionization energy. An electron just barely ionized from an atom exists in a state of zero energy (excluding its rest energy). In the hydrogen atom, the electron reaches zero energy, when its quantum number n approaches infinity. It thus requires an energy of 13.6 eV to raise a hydrogen electron from the ground state to the lowest ionized state. The constant E_0, in Bohr's equation for energy quantization, then, denotes simply the ionization energy of hydrogen.

When Bohr's theory of the hydrogen atom appeared in 1913, the scientific world was astonished. To a generation educated in a prequantum era, the rising prominence of the quantum in the atomic domain was, perhaps, both confusing and captivating. Rutherford himself reacted to the Bohr theory with some mixed feelings: "Your ideas . . . are very ingenious and seem to work out well; but the mixture of Planck's ideas with the old mechanics makes it very diffcult to form a physical idea of what is the basis of it all."

QUESTIONS

19–1. Obtain the numerical value of the frequency of the fifth line in the visible spectrum of hydrogen as a fractional multiple of the constant K.

Answer: $\dfrac{45}{196}$

19–2. Comment on the value of an empirical formula. What role does the empirical approach play in your own discipline?

19–3. Cathode rays, like light rays, cast shadows. If a magnet is used to distinguish them, what fundamental differences would you observe in the two types of rays?

19–4. Thomson was not able to measure the charge, or the mass, of the electron by itself. He was only able to measure the ratio.

From his measurement of the charge-to-mass ratio for ionized hydrogen, he deduced that an electron must be some 2000 times lighter than a hydrogen nucleus. What assumption did he have to make for such a deduction? Justify the assumption.

19–5. The gold foil in Rutherford's alpha-particle-scattering experiment was about 4×10^{-5} cm thick. If the interatomic distance is of the order of 10^{-8} cm, how many layers of gold atoms did the alpha particles have to transverse?

Answer: 4000

19–6. Bohr borrowed the following equation from the physics of Newton and Maxwell:

(1) $\qquad F = ma \qquad$ or $\qquad \dfrac{e^2}{r^2} = m\left(\dfrac{v^2}{r}\right)$

(2) $\qquad \frac{1}{2}\, mv^2 - \dfrac{e^2}{r} = E$

He then imposed the quantum condition:

$$mvr = n\hbar$$

where $\qquad\qquad \hbar = h/2\pi.$

From these equations he showed that the orbit of the electron must be quantized according to

$$r = n^2 r_0, \quad n = 1, 2, 3, \ldots$$

where $\qquad\qquad r_0 = \hbar^2/me^2.$

(a) Show that

$$E = \dfrac{-E_0}{n^2}$$

where

$$E_0 = \dfrac{me^4}{2\hbar^2}$$

(b) Given that

$$\dfrac{e^2}{\hbar c} = \dfrac{1}{137}$$

and the rest energy of the electron, mc^2, is 5.1×10^5 eV, obtain the numerical value of E_0

(c) If the numerical value of $\hbar/mc$ is 3.86×10^{-11} cm, what is the numerical value of the radius of the first Bohr orbit?

Light from Hydrogen_____

19–7. Calculate the frequency of red light emitted from the hydrogen atom. For Planck's constant use the value of $h = 4.15 \times 10^{-15}$ eV-sec.

19–8. Photons having frequencies higher than about 7×10^{14} vibrations per second are not visible. If an electron in the hydrogen atom makes a downward transition from the $n = 2$ state to the $n = 1$ state, what is the energy of the photon? Would it be visible?

19–9. Would a hydrogen atom in the ground state absorb visible light? Explain.

Answer: no

19–10. (a) If a gas of hydrogen atoms is bombarded with photons of 12.1 electron volts of energy, what state would the atoms be excited to?

Answer: $n = 3$

(b) Three groups of photons are observed when these atoms return to their ground states. From an energy level diagram, find the energy of each group of photons.

Answer: 10.2 eV, 12.1 eV, 1.9 eV

19–11. Explain Balmer's empirical formula in terms of the Bohr model of the atom.

19–12. Bohr treated the orbiting electron in the hydrogen atom nonrelativistically, using for the mass of the orbiting electron its rest mass.

(a) Calculate the velocity of the electron in the $n = 1$ orbit from $mvr = \hbar$, where $r = \hbar^2/me^2$. (Use $\hbar c/e^2 = 137$.)

Answer: $c/137$

(b) Is Bohr's nonrelativistic model justified?

SUGGESTIONS FOR FURTHER READING

MARION, J. B. *A Universe of Physics*. New York: John Wiley & Sons, Inc., 1970. (Section 5.1, "J. J. Thomson: Cathode Rays"; Section 6.2, "J. J. Balmer: Note on Spectral Lines of Hydogen"; Section 6.3, "N. Bohr: On the Constitution of Atoms and Molecules.")

Physics: The Fabric of Reality_____

20

The Periodic Table of the Elements

One of the earliest confirmations of the Bohr theory of the atom was provided by H. G. J. Moseley. He demonstrated that the most intense X-ray spectral lines emitted by the elements are described by the Bohr theory. He further established that the elements, arranged according to their X-ray spectral lines, followed the periodicity of the elements discovered by Mendeleyev.

The Periodicity of the Elements

It all began with preparations for his book entitled *Principles of Chemistry,* published in 1868. Struck by a lack of classification of the chemical elements. Dmitri Mendeleyev (1834–1907) set out to gather every piece of information about the chemical and physical properties of the sixty-five known elements. This work resulted in his discovery of a periodic law of the elements: *The properties of the elements are a periodic function of their atomic weights.*

Mendeleyev then proceeded to construct a table consisting of rows and columns. He placed the elements in the horizontal row in the order of increasing atomic weight. The second row of elements was established to repeat the chemical and physical properties of the elements in the first row. As the rows were thus filled the resulting columns were comprised of the elements of similar properties. Mendeleyev did not know why such periodic relationships existed among the elements. Nevertheless, he recognized these relationships to be of fundamental significance and went on to predict the existence of certain elements to fill the empty spaces in his table of the elements. These elements were all discovered subsequently.

The Atomic Number

In 1913 a discovery of fundamental importance in the understanding of the periodic nature of the elements was made by H. G. J. Moseley (1887–1915). He bombarded targets of different metals with an energetic electron beam, and observed the photons of high frequency that were emitted by the target atoms upon de-exitation. These photons were characteristic of the struck atoms, and were called characteristic X rays. He noticed that the frequencies of the characteristic X rays of the elements increased with the atomic weights paralleling the ordering of the elements in Mendeleyev's periodic table. He thus suspected that there might exist a physical basis for the periodicity of the elements.

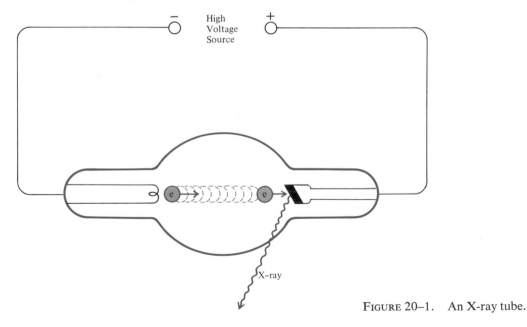

FIGURE 20–1.   An X-ray tube.

An atom may emit several characteristic X-ray spectral lines with varying intensities. Moseley noticed that the observed frequency of the most prominent (intense) X-ray spectral line of an element could be expressed as

$$f = K(Z - 1)^2 \left(\frac{1}{1^2} - \frac{1}{2^2} \right)$$

where K is a constant common to all elements and Z is an integer characteristic of the element. He further noticed that Z increased by 1 from one element to the next in the periodic table. He thus concluded that a fundamental number may be assigned to an element. This fundamental number for aluminium, for instance, was found to be 13, and since aluminium is the thirteenth element in the periodic chart, Z was called the atomic number.

The physical significance of the atomic number is provided by the Bohr model of the atom. There is a spectral line of hydrogen which is expressed by the Bohr theory as

$$f = KZ^2 \left(\frac{1}{1^2} - \frac{1}{2^2} \right)$$

where Z represents the nuclear charge number of the hydrogen nucleus. For hydrogen the nuclear charge number is unity. Its one unit of positive charge is then balanced by one unit of negative charge of the electron. Thus, in the Bohr model, the expression Z represents an effective positive charge that the electron "sees" on the nucleus. The expression $(Z - 1)$, in Moseley's formula, may then be interpreted as the effective positive charge that would be "seen" by the electron making the quantum jump to emit the observed X-ray spectral line. Since it was the number Z that changed with the element, it was reasonable to associate the quantity Z with the nuclear charge number rather than the quantity $(Z - 1)$. The effective charge $(Z - 1)$ was then pictured as arising from the shielding of the nucleus by the innermost electron, so that the electron jumping from the $n = 2$ orbit to the $n = 1$ orbit would "see" only $(Z - 1)$ units of positive charge.

Moseley thus established that the periodicity of the elements is a function of the atomic number rather than of the atomic weight, as thought by Mendeleyev. He further argued that the physical basis for the periodic table must then be found in the arrangement of the electrons in the atom, since the atomic number also determines the number of electrons in a normal atom.

The Periodic Table of the Elements_____

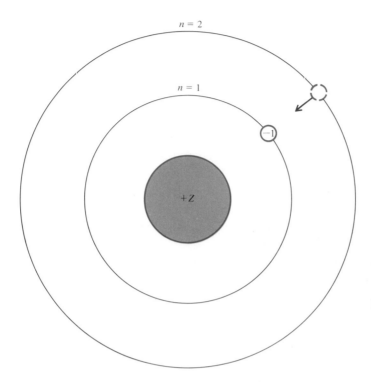

FIGURE 20–2. An effective charge, $(Z - 1)$, is "seen" by the electron jumping from the $n = 2$ orbit to the $n = 1$ orbit.

States of Motion for Atomic Electrons

Before one can know how the electrons are arranged in the atom, one must first know how to specify the states of motion for the atomic electrons. In the original Bohr model of the atom, the electron may exist in any allowed circular orbit. A circular orbit was specified by an integer, ordering orbits by increasing radius, and this integer was referred to as the principal quantum number. An electron existing in an nth state of motion, for example, was pictured as describing an nth circular orbit. The states of motion for the atomic electron were thus specified by the values of the principal quantum number n, which assumed integral values $n = 1, 2, 3, \ldots$.

As observations of atomic spectra grew more refined and sophisticated, it became obvious that the principal quantum number alone was not sufficient to specify the quantum state of an atomic electron. This led A. Sommerfeld, in 1915, to introduce the concept of elliptical orbits.

He found that in order for the Bohr model to give results in agreement with the observed atomic spectra, there had to be assigned an n possible number of shapes for an nth orbit. An electron in the $n = 3$ orbit, for example, could describe three different shapes of orbit; a circular, an elliptical, or a nearly straight line, orbit. The different shapes of an orbit were specified, again, by a set of integers called the angular momentum quantum number, l. The allowed values were $l = 0,1,2,3,..., n - 1$. The higher the value of the angular momentum quantum number, the more circular was the orbit.

When atoms were subjected to the influence of a magnetic field, the atomic spectra were observed to be still more complex. It thus became necessary to introduce a third quantum number, called the magnetic quantum number, m. The magnetic quantum number specifies the orientation of the plane of orbit relative to the direction of the magnetic field. A given shape of an orbit specified by an angular momentum quantum number l may assume a $(2l + 1)$ number of possible orientations. Symbolically, for a given l, the magnetic quantum number assumes values $m = 0, \pm 1, \pm 2,..., \pm l$.

The orbital motion of an atomic electron is completely specified by the three quantum numbers n, l, and m. The ground state electron of a hydrogen atom, for example, has quantum numbers $n = 1$, $l = 0$, and $m = 0$.

However, it was later discovered that an atomic electron executes spinning motion about its own axis as well as orbital motion. The motion of an atomic electron is thus reminiscent of the motion of the earth. The

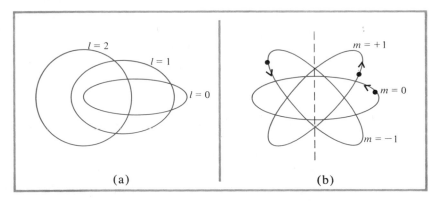

FIGURE 20–3.   (a) Three possible shapes of the $n = 3$ orbit; (b) three possible orientations of an elliptical orbit.

The Periodic Table of the Elements⎯⎯⎯⎯⎯⎯⎯⎯⎯⎯⎯⎯⎯⎯⎯⎯

electron's spinning motion is specified by a quantum number called the spin quantum number s. The electron spins only in two possible directions. If it rotates from left to right, it is said to spin up and is assigned a spin quantum number $s = +\frac{1}{2}$. If it spins in the opposite direction, it is said to spin down, and is assigned a value $s = -\frac{1}{2}$.

These four quantum numbers are found to be sufficient to specifiy the quantum state of motion for an atomic electron. The question as to how the electrons are arranged in an atom, however, requires the discovery of a principle that governs these quantum states of motion.

The Exclusion Principle

That the electrons in an atom do not all exist in the innermost first orbit was generally accepted from the beginning. The Bohr model predicted that the first orbit of the hydrogen atom would be greater than that of any other atom by a factor of the latter's atomic number. If all atomic electrons existed in the first orbit, the hydrogen atom would then be some eight times as great in size as the oxygen atom, and some ninety-two times as great as the uranium atom. Observation, however, indicated that all atoms were roughly equal in size. It was thus clear that the electrons in an atom were somehow stacked up.

W. Pauli discovered in 1924 that no two electrons in the same atom could exist in the same quantum state of motion. This important behavior of the electron is referred to as the exclusion principle. It then follows directly from the exclusion principle that there can be no more than two quantum states available to an electron in the first ($n = 1$) orbit. Therefore, no more than two electrons can be accommodated in the first orbit. There are eight available quantum states in the second ($n = 2$) orbit, and the second orbit can accommodate up to eight electrons. The electrons in an atom thus fill up the lowest available quantum states according to Pauli's exclusion principle.

Now the periodicity of the elements may be explained in terms of atomic structure. Compare the elements lithium and sodium, for example. Both belong to the same column in the periodic table and thus have similar chemical and physical properties. Lithium has three electrons and sodium has eleven electrons. The outermost electron of the lithium atom must go to the second ($n = 2$) orbit, as the first orbit is filled with its

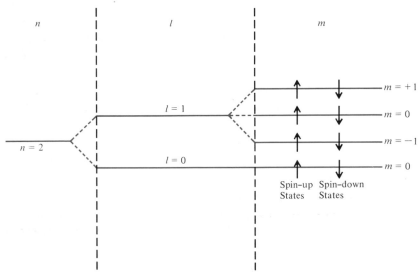

FIGURE 20–4. An explicit display of the available quantum states in the $n = 2$ orbit.

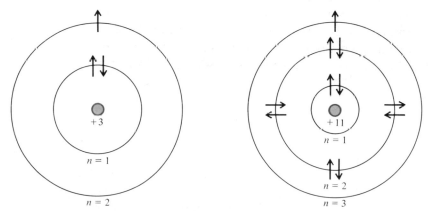

FIGURE 20–5.   The electron configurations of lithium and sodium.

maximum of two electrons. The outermost electron of the sodium atom goes to the third ($n = 3$) orbit, with the first two orbits filled with the other ten electrons. Thus, in both lithium and sodium, the outermost electron is a single electron existing outside a filled atomic core. The chemical and physical properties of both elements are largely determined by the single outermost electron, and, therefore, are similar.

The Periodic Table of the Elements

211

We have thus established a physical basis for the periodicity of the elements. We have done so by modifying and extending the Bohr model of the atom, and by imposing on the atomic electrons Pauli's exclusion principle. The Bohr model, however, could not cope with the experimental results of increasing refinement and sophistication. Even on aesthetic grounds, the Bohr theory—its conceptual simplicity notwithstanding—drew increasing objection as a noncoherent theory of ad hoc nature. The atom had simply to be comprehended in a more coherent way.

QUESTIONS

20–1. One could say that Mendeleyev's periodic law for the elements played a role in our understanding of the elements similar to that played by Balmer's empirical formula in our understanding of the hydrogen atom. Explain.

20–2. When an electron in the hydrogen atom makes a transition from the $n = 2$ state to the $n = 1$ state, a photon carrying an energy of 10.2 eV is emitted. Calculate the energy of the most intense X-ray photon that Moseley would have observed upon bombarding an aluminium target with energetic electrons. (For aluminium, $Z = 13$).

Answer: 1,468.8 eV

20–3. X rays and visible light are both electromagnetic radiation. The difference lies in their frequencies, or energies. The range of energy carried by visible light is 1.5 eV to 3.0 eV. X rays, on the other hand, have energies in the thousands of electron volts. Explain why X-ray spectra are associated with "inner shells" (or orbits) and visible spectra with "outer shells."

20–4. Show explicitly why the $n = 1$ state in an atom can accommodate no more than two electrons.

20–5. Recall that Dirac applied Pauli's exclusion principle to the electrons in the negative-energy ocean. Was Dirac really justified in applying the exclusion principle to these unphysical electrons? Explain.

20–6. If the Pauli exclusion principle were not in operation, how would our physical world be different?

20–7. Explain why helium ($Z = 2$) and neon ($Z = 10$) have similar chemical properties.

20–8. Insofar as the excitation of the outermost electron is concerned, the sodium atom ($Z = 11$) can be viewed as a one-electron atom, that is, as hydrogenlike. Explain.

20–9. A sodium arc gives off yellow light. Common salt (NaCl), thrown into a flame, burns bright yellow. Explain qualitatively the origin of the yellow light.

SUGGESTIONS FOR FURTHER READING

GAMOW, G. "The Exclusion Principle," *Scientific American* (July 1959).
WEIDNER, R. T., and R. L. SELLS. *Elementary Modern Physics*. Boston: Allyn and Bacon, Inc., 1960. (Chapter 7: "X-ray Spectra.")

The Periodic Table of the Elements_____

21

The Wave Nature of the Electron

The idea of an orbiting electron is central to the Bohr model of the atom. As a physical entity executing orbital motion, the atomic electron is regarded as a particle, occupying an infinitesimal region of space at any instant of time. As such the atomic electron follows only certain quantized orbits. The Bohr theory, however, provides no justification as to why atomic electrons choose to exist only in those allowed orbits. Nor does the theory describe the mechanism whereby an electron makes a quantum jump between the orbits. Thus the Bohr theory is fraught with inherent weaknesses, of which the view of the electron as a particle executing incessant orbital motion is perhaps the most serious.

The Wave Nature of the Electron

The idea that the electron may be viewed as a wave originated with L. de Broglie in 1924. As he studied carefully Einstein's revolutionary idea of the duality of light—that light must be regarded sometimes as a particle and sometimes as a wave—de Broglie wondered if this notion of duality for light could not perhaps be generalized to matter as well. But if matter is to be regarded as having wave properties, the wavelength associated with matter must be known.

De Broglie sought the answer in the theory of relativity. Einstein's relativistic energy–momentum relationship takes, for the photon, a very simple form:

$$E = pc$$

where E and p denote the energy and momentum of a photon, and c is the speed of light. This relationship is of particular significance in that it defines the ratio of energy to momentum to be a universal constant, namely, the speed of light.

Now the energy of a photon is given by its frequency f, or, equivalently, by its wavelength λ.

$$E = hf = h\left(\frac{c}{\lambda}\right)$$

where h is Planck's constant. The momentum of a photon may then be expressed as

$$p = \frac{E}{c}$$

or

$$p = \frac{h}{\lambda}$$

The most interesting aspect of this relationship is that the product of the momentum and the wavelength of a photon defines a universal constant, namely, Planck's constant. A deep, inseparable link between momentum and wavelength is thus suggested. Capitalizing on it, de Broglie postulated that not only light but matter also—thus everything in the universe— obey such a relationship. Specifically, the wavelength of matter would be determined by its momentum.

$$\text{wavelength associated with matter} = \frac{\text{Planck's constant}}{\text{momentum}}$$

or

$$\lambda = \frac{h}{mv}$$

De Broglie applied his wave hypothesis to the Bohr model of the hydrogen atom. The Bohr model predicts the circumference of the first orbit to be about 3.3×10^{-8} cm and the velocity of the orbiting electron to be about $c/137$. The wavelength of such an electron, according to de Broglie's hypothesis, comes out to be

$$\lambda = \frac{h}{mv} = 3.3 \times 10^{-8}\,\text{cm}$$

An obvious interpretation of the de Broglie wavelength of the electron is that the circumference of the first Bohr orbit corresponds to exactly one wavelength of the electron.

More generally, de Broglie assumed that if the electron wave is to exist in an atom, it must exist as a standing wave. (A macroscopic example of a standing wave is the wave form assumed by a vibrating string when both ends are fixed.) A standing wave would be achieved in an atom, if the circumference of an orbit is an integral multiple of the de Broglie wavelength. That is,

$$2\pi r = n\lambda = n\left(\frac{h}{p}\right), \quad n = 1, 2, 3, \ldots$$

or

$$pr = n\left(\frac{h}{2\pi}\right)$$

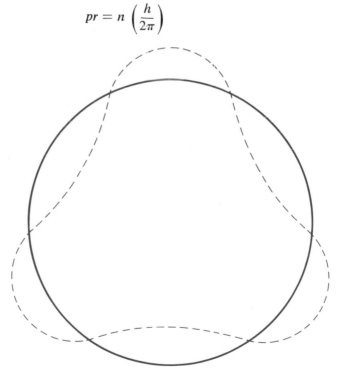

FIGURE 21–1. A de Broglie wave in the third Bohr orbit.

The Wave Nature of the Electron

The last expression is exactly the quantum hypothesis which Bohr introduced in his atomic theory as an ad hoc postulate.

It thus appears that the mystery of quantized orbits in the Bohr theory may be rooted in the wave nature of the electron, for an electron viewed as a wave cannot exist just anywhere within an atom. Rather, it can exist only at those regions where it can achieve a standing wave. The de Broglie hypothesis of electron waves thus carries a great deal of appeal. It may, indeed, turn out to be an essential ingredient in the construction of a fundamental atomic theory.

Electron Waves—An Experimental Proof

If it is to qualify as a wave, an electron must display certain properties that are common to all types of waves. Since interference and diffraction are distinct wave phenomena, electrons must, at least, display interference and diffraction effects.

Interference results when waves superpose. Consider, for the sake of simplicity, the interference of two similar waves. If the crest of one wave coincides with the crest of another, the amplitude of the resulting wave is twice as large as that of the individual wave. The wavelength and frequency of the new wave, however, remain the same as those of the individual waves. Such an interference is said to be constructive. If, on the other hand, the waves combine such that the crest of one wave coincides with the trough of another, a total cancelation of waves results, which is

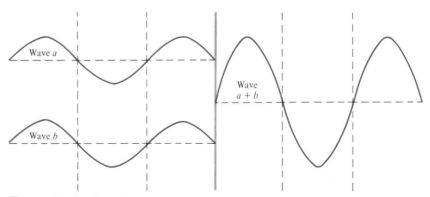

FIGURE 21–2.   Interference of two similar waves.

said to be a destructive interference. There are, of course, an infinite number of ways that two similar waves can interfere, leading to an infinite number of possibilities for the superposed wave.

Another exclusive property of waves, referred to as diffraction, is that waves can bend around the corners of an obstacle. The image of the edge of a razor blade, for example, is not sharp like the edge itself, but consists of a series of dark and bright fringes as a result of the diffraction of light by the razor edge. Referred to as a diffraction pattern, the series of dark and bright fringes in the image arises from the interference of the light waves that bend around the edge of the razor blade, spreading in all directions.

Diffraction patterns are most readily demonstrated when light is directed through a narow rectangular slit and the image of the slit is projected on a screen. If the width of the slit is much greater than the wavelength of light, the image is a simple geometrical replication of the slit itself. If, on the other hand, the slit width is comparable to the wavelength of light, the screen displays a diffraction pattern. The diffraction pattern

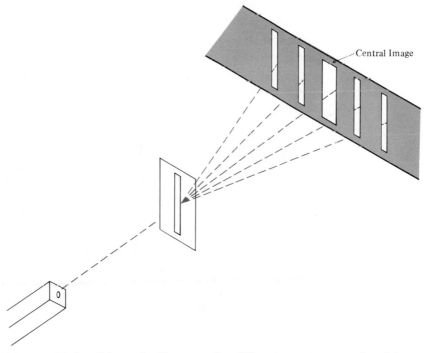

FIGURE 21–3. Schematic diagram of a diffraction pattern produced by a laser beam passing through a narrow rectangular slit.

The Wave Nature of the Electron_____

consists of a bright central image wider than the slit itself, and, on either side, a series of secondary images separated by dark regions. These secondary fringes decrease in brightness as they get farther away from the central image.

Now if electrons can be demonstrated to produce a diffraction pattern, such would constitute a compelling proof of the wave nature of the electron. De Broglie himself suggested that electrons, bounced off the surface of a crystal, might produce a diffraction pattern. The suggestion derives from the fact that the wavelength associated with a readily available electron beam is comparable to the interatomic spacings in a crystal. Such electrons would behave very much like X rays, which also have wavelengths comparable to the interatomic spacings in a crystal. When X rays shine on a crystal, the atoms scatter the X rays, causing them to interfere so as to produce a diffraction pattern. In fact, the interatomic spacings of crystals are deduced from the observed diffraction patterns of X rays.

It so happened that C. J. Davisson at the Bell telephone Laboratories had been scattering electrons from metals since 1919, unaware of de Broglie's work. In April of 1925, as he and Germer were bouncing electrons off a nickel sample in a high vacuum, an accident broke the vacuum and wrecked their apparatus, so that their nickel sample was oxidized by the inrushing air. As a way of cleaning the nickel surface, the sample was subjected to prolonged heating. When they resumed their experiment with the heat-treated nickel target, they observed a totally unexpected, unsuspected result—an X-ray diffraction pattern. But there

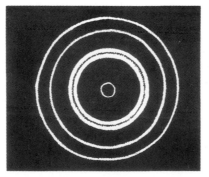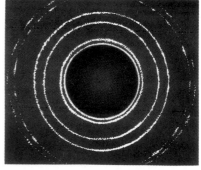

FIGURE 21–4. Comparison of diffraction patterns from an aluminum foil (produced by (*left*) X rays and (*right*) electrons, both having equal wavelengths. (After the Physical Science Study Committee film *Matter Waves*.)

was no X ray present. Thus the diffraction pattern they observed had to be attributed to the electrons.

More specifically, Davisson and Germer used an electron beam having a kinetic energy of 54 electron volts. These electrons had a de Broglie wavelength of 1.67×10^{-8} cm. The heat treatment of the nickel target had caused the nickel atoms to align themselves in a regular, crystalline arrangement. These atoms were about 2.1×10^{-8} cm apart, so that their spacings were comparable to the de Broglie wavelength of the electron beam. A general theory of wave motion then predicted that an electron diffraction peak would occur at 50 degrees from the direction of the incident electrons. An electron detector, counting the scattered electrons as a function of the angle, registered a peak counting rate exactly at the predicted angle. The de Broglie hypothesis of matter waves was thus confirmed.

The Wave-Particle Dilemma

A wave is specified by a wavelength, or equivalently by a frequency. If it is to have a precise wavelength, a wave must consist of crests and troughs that repeat themselves an infinite number of times. But such a wave would have an infinite extension in space. An ideal wave may thus be characterized by an infinite spatial extension.

An ideal particle, on the other hand, is a physical construct which has no size associated with it. As it has a zero spatial extension, a particle must be precisely localizable in space.

An ideal wave and an ideal particle are mutually contradictory concepts. It is for this reason that the wave-particle duality of light and matter is so disturbing a notion. And yet pysical phenomena require just such a dualism. Light, for instance, must be described as a wave when it interferes and diffracts, but as a particle when it knocks out an electron from a metal surface. Similarly, an electron must be described as a charged particle when it is subjected to an electric field, but as a wave when it bounces off a crystal. The dual nature of a physical entity thus seems inescapable.

The wave-particle dilemma becomes even more transparent in the case of a free electron at rest. Viewed as an ideal particle, the electron would occupy a well-defined point in space. But calculate its de Broglie wavelength—it is infinite! Viewed as a wave, the electron must thus have an infinite extension in space. But an electron cannot, at once, be localized and extended infinitely in space. Such a situation would clearly be nonsense.

Resolution of the Wave-Particle Dilemma

Let us concede that a wave of infinite spatial extension cannot represent a localized particle. In the hope that the wave-particle dilemma may be resolved in a compromise between an ideal wave and an ideal particle, let us phrase the question as follows: Under what conditions would a wave become localized, and what would be the implication of a localized wave for a physical object that it is presumed to describe?

To begin with, consider two waves of the same amplitude but of slightly different frequency. If they interfere, the resulting wave would have a varying amplitude. In the case of sound waves, such a condition arises when, for example, two adjacent piano keys are struck simultaneously. The varying amplitude gives rise to variations in loudness which are called beats. Beats represent localized wave packets. Ideal beats, however, are still infinite in spatial extension, as their variations in loudness must repeat indefinitely in order to achieve a precise beat frequency.

A localization of a wave, however, may be achieved if many monochromatic waves are superposed. Suppose all waves lying within a wavelength band of $\Delta\lambda$ are superposed on a wave whose wavelength is λ. If all such individual waves are set in phase at a particular point in space, they would interfere constructively, giving rise to a large resultant amplitude at

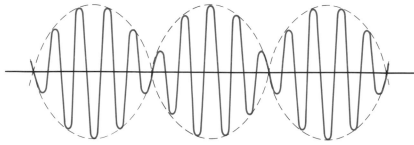

FIGURE 21–5. Superposition of two waves of slightly different wavelengths produces a beat.

Physics: The Fabric of Reality

that point. On either side of the central point, the waves would become increasingly out of phase, so that the amplitude of the superposed wave would rapidly approach zero toward the ends of a spatial extension Δx. The spatial extention Δx of the resulting wave packet is determined by the spread in wavelength $\Delta\lambda$ of all the superposed waves:

$$\Delta\lambda\Delta x \geqslant \lambda^2$$

where λ is the wavelength of the original wave. Thus, if the spread in wavelength is small, the spatial extension of the wave packet becomes large. On the other hand, if the wave packet is to be localized to a small region, waves over a wider range of wavelengths must be superposed.

This peculiar feature of wave motion holds a startling implication for a microscopic object. According to the de Broglie hypothesis, an object with a well-defined momentum would have a well-defined wavelength, and would thus have an infinite extension in space. A single de Broglie wave, therefore, cannot describe an object that is spatially localized. A localized object must then be represented by a superposition of de Broglie waves with different wavelengths. A spread in the de Broglie wavelength, however, would imply a spread in the momentum of the object, because momentum and wavelength are inseparably linked. Mathematically, the relationship is given by

$$\Delta p = \frac{h\,\Delta\lambda}{\lambda^2}$$

where Δp and $\Delta\lambda$ denote the spread in momentum and wavelength, respectively.

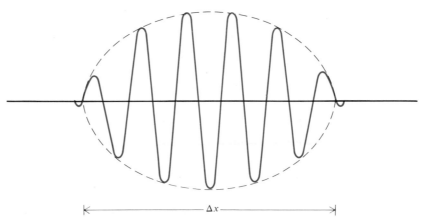

FIGURE 21–6.　Superposition of many waves, producing a wave packet.

The Wave Nature of the Electron

But a spread in wavelength is also related to a spread in the spatial extension of a wave packet. Thus a relationship can be established between Δp and Δx, or between the spread in the momentum of an object and the size of a wave packet describing the object. If we compute the product of Δx and Δp, we obtain the result

$$\Delta x \Delta p \geqslant h$$

which may be interpreted as follows: If an object is to be described by a de Broglie wave packet, the size of the wave packet Δx must correspond to an *uncertainty* in the position of the object. Thus the position of an object cannot be precisely determined. But neither can the momentum of an object be determined precisely, as a wave packet cannot be constructed without first allowing the momentum to have a spread, i.e., to be uncertain by Δp. Thus if a physical object is to achieve a measure of localization in space at all, it must possess an inherent uncertainty in its momentum. The wave-particle dualism thus leads to this strange result, which W. Heisenberg proposed in 1927 as a fundamental law of nature. Now referred to as the principle of uncertainty, the principle may be stated: *There exists a fundamental limit on the ultimate precision with which the position and momentum of an object may be measured simultaneously.*

The Principle of Uncertainty

One's immediate reaction to the uncertainty principle is that somehow such a principle could not really be a law of the universe. Heisenberg himself wondered whether the reciprocal limitation of precision in the simultaneous measurements of position and momentum is merely a restriction imposed by mathematics or whether it is a reflection of a more profound scheme of things.

Heisenberg did concede that one can indeed determine the position of an object to any desired degree of accuracy. But the notion of position, he continued, is meaningless apart from an experiment designed to measure it. A measurement of the position, however, would be impossible without disturbing the object, and a disturbance would introduce an uncertainty in the momentum of the object. Thus, absolute precision in the simultaneous measurement of position and momentum cannot be attained, even in principle.

A thought experiment to illustrate Heisenberg's argument was suggested by Bohr in 1928. The experiment consisted in observing an electron by illuminating it with radiation and by detecting the scattered radiation with a microscope. There are two effects involved in such an observation. The first effect has to do with the wave nature of light. Because of diffraction, two nearby spots, for example, cannot be seen as two separate spots unless their distance of separation is of the order of, or greater than, the wavelength of the illuminating light. The second effect has to do with the quantum nature of light; namely, a photon delivers part of its momentum to the electron from which it scatters, thus disturbing the electron.

A precise determination of the position of an electron would, therefore, require that light of a very short wavelength be used. A short wavelength, however, implies a large momentum, so that the impact of the photon would change the electron's momentum abruptly. In fact, a photon would

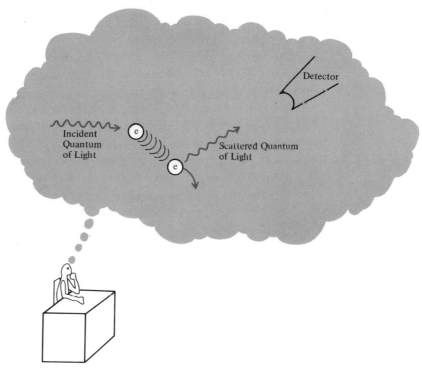

FIGURE 21–7.   Bohr's thought experiment.

The Wave Nature of the Electron

deliver to the electron a momentum comparable to its own momentum, namely, h/λ. But the position of the electron can be determined only within the wavelength λ of light. Thus one encounters a situation where one can determine the position of an electron more precisely only by disturbing it more. Or reduce the disturbance on the electron, and the measurement of the electron's position suffers a loss of precision.

Specifically, if an electron is to be located within a spatial region of $\Delta x = \lambda$, it must necessarily suffer a disturbance, so that its momentum would be uncertain by an amount imparted by the photon, i.e., $\Delta p = h/\lambda$. If we now calculate the product of Δx and Δp, we again obtain the result, $\Delta x \, \Delta p = h$. The reciprocal limitation of precisions in the simultaneous measurements of position and momentum is thus inescapable also on physical grounds.

The uncertainty principle is not limited to the simultaneous measurements of position and momentum. The theory of relativity dictates that if an uncertainty relation exists for position and momentum, an uncertainty relation must also exist for time and energy. This is so because, in relativity, space is related to time in a similar manner as momentum is to energy. The similarity is such that momentum and energy are referred to as the space and time components of the relativistic four-momentum. Relativity would thus limit precision on the simultaneous measurements of time and energy, also. Furthermore, the uncertainty relation for time and energy would be $\Delta t \, \Delta E \geqslant h$.

The uncertainty principle has profound implications. It first of all recognizes the fact that, in the quantum world, there exists an unavoidable interaction between the observer and the observed. The observer, in fact, becomes an integral part of the observed phenomenon, so that a quantum phenomenon per se does not exist even in principle. The uncertainty principle, also, implies that a physical theory must be founded ultimately on observable concepts. Because neither an ideal wave nor an ideal particle can be observed, they cannot describe a physical object. It is thus legitimate to reject the concept of ideal wave and ideal particle. But in so doing, one must also abandon the notion that the position and the momentum of a physical entity can be observed simultaneously with infinite precision. The reciprocal limitation on precision, as revealed by the uncertainty principle, is thus an inescapable consequence of the wave-particle duality. As such the uncertainty principle not only makes sense out of the wave-particle duality but renders beautifully comprehensible the quantum world itself.

QUESTIONS

21-1. Consider two electrons moving at nonrelativistic velocities, i.e., $v << c$. If one electron has a velocity twice as large as the other, how would their respective de Broglie wavelengths compare?

Answer: $\lambda_1 = \lambda_2/2$

21-2. Do macroscopic objects also have de Broglie wavelengths associated with them? If so, why are they not observed?

21-3. Show that in terms of the total energy E and the rest energy E_0, the de Broglie wavelength of an electron is given by $\lambda = hc/\sqrt{E^2 - E_0^2}$.

21-4. The de Broglie wavelength of an electron at rest is infinite. Viewed from a moving frame of reference, the electron would not be at rest but would be in motion, so that its de Broglie wavelength would be finite. Argue why the de Broglie wave associated with the electron would, in either case, have an infinite spatial extension.

21-5. Suppose many waves are superposed to form a wave packet whose spatial extension is Δx. If λ is the wavelength of the original wave to which waves of slightly different wavelengths are added, show that the shortest wavelength in the group of superposed waves is given by $\lambda (\Delta x \quad \lambda)/\Delta x$.

21-6. Given that $\Delta x \geqslant \lambda^2/\Delta\lambda$ and $\Delta p = h \ \Delta\lambda/\lambda^2$, derive the uncertainty relation for position and momentum.

21-7. It can be shown by use of the uncertainty principle that the radius of the first orbit of the hydrogen atom should be 0.5×10^{-8} cm, in agreement with the prediction of the Bohr model. Explain qualitatively why the electron could not orbit closer to the nucleus.

21-8. The uncertainty relation for time and energy can be interpreted as suggesting that the conservation of energy can be violated by an amount ΔE if the violation lasts no longer than a time interval Δt given by the uncertainty relation. A process which briefly violates energy conservation is called a virtual process. An example of a virtual process is an electron emitting a photon which, in turn, transforms into a pair of electron and positron.

The virtual pair, upon annihilation, turns into a virtual photon, which is subsequently absorbed by the original electron. The entire process takes place within a time limit governed by the uncertainty principle. Could an electron, then, also create a pair of elephant and antielephant? Explain.

21–9. Comment on the claim that the observer is a part of the phenomenon.

SUGGESTIONS FOR FURTHER READING

COOPER, L. N. *An Introduction to the Meaning and Structure of Physics,* short edition. New York: Harper & Row, Publishers, 1970. (Chapter 32: "The Electron as a Wave.")

GAMOW, G. "The Principle of Uncertainty," *Scientific American* (January 1958).

JAMMER, M. *The Conceptual Development of Quantum Mechanics.* New York: McGraw-Hill Book Company, 1966. (Section 5.3: "The Rise of Wave Mechanics.")

22

The Schrödinger Theory of the Atom

Duality conforms to reality in the quantum world. The wave aspects of matter are as legitimate as the particle aspects. In fact, the theory that succeeds in describing the atom in a coherent manner is founded upon a physcial law governing the behavior of matter waves. Such a law, discovered by Erwin Schrödinger in 1926, is referred to as the Schrödinger wave equation.

Waves on a String

Consider a string fixed at both ends. Set the string in vibration, and it would vibrate only at certain frequencies. These are called the natural, or resonant frequencies, of the string, and may be deduced from the shapes of the wave form that the string may assume. The wave forms must satisfy the condition that the amplitude of the wave must be zero at the fixed ends of the string. Such wave forms are described by a set of wavelengths,

$$\lambda_n = \frac{2L}{n}$$

where n specifies the nth wave form, and L is the length of the string.

Now to each shape of the wave form there corresponds a frequency of

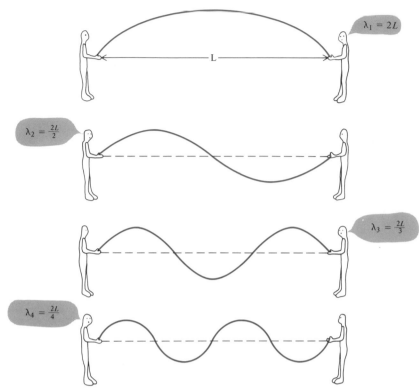

$$\lambda_1 = 2L$$

$$\lambda_2 = \frac{2L}{2}$$

$$\lambda_3 = \frac{2L}{3}$$

$$\lambda_4 = \frac{2L}{4}$$

FIGURE 22–1. Allowed wave forms on a string when both ends are fixed.

vibration, so that the natural frequencies also constitute a discrete set. Specifically, the discrete set of frequencies is given by

$$f_n = nf_1, \quad n = 1, 2, 3, \ldots$$

where f_1 is the fundamental frequency, and f_n is an nth harmonic. The frequency of a string with fixed ends is thus quantized.

The de Broglie Wave of a Bound Electron

Consider now an electron constrained to move along the x axis between two impenetrable walls. Let these walls be separated by a distance L. Inside such a linear "box" the electron would experience no force. At the walls, on the other hand, it would experience an infinite force, thus being bounced back into the box instantaneously with no loss of energy. The question of interest is, "What type of motion would such an electron execute?"

Physics: The Fabric of Reality

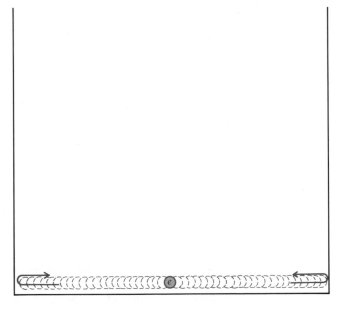

FIGURE 22–2. An electron confined in a "linear" box with impenetrable walls.

If the motion is viewed strictly as a particle behavior, such an electron would be either in a state of rest or in a state of uniform motion. The energy states would thus form a continuum.

If the electron is viewed as a wave, on the other hand, an entirely different situation emerges. The question reduces to finding out the allowed shapes of the de Broglie wave associated with the electron. Now the conditions imposed on the electron wave are similar to those imposed on a vibrating string whose ends are fixed. We know that for such a string only standing waves exist. It is therefore unlikely that a nonstanding de Broglie wave would represent a state of motion for the electron. The de Broglie wave forms associated with the electron are more likely to be specified by a discrete set of wavelengths such as

$$\lambda_1 = \left(\frac{2}{1}\right) L$$

$$\lambda_2 = \left(\frac{2}{2}\right) L$$

$$\lambda_3 = \left(\frac{2}{3}\right) L$$

. . . . . .

$$\lambda_n = \left(\frac{2}{n}\right) L$$

where n is an integer specifying the nth wave form.

The Schrödinger Theory of the Atom

Now according to the de Broglie hypothesis, the wavelength and the momentum of a particle are related in an inseparable way. The determination of one fixes the value of the other uniquely by $p = h/\lambda$. Thus a discrete set of wavelengths for an electron confined in a linear box would imply a corresponding discrete set of momenta for the electron; namely, $p_n = h/\lambda_n$. But a discrete set of values for momentum, in turn, leads to a discrete set of states for the motion of the electron.

The states of motion are commonly specified by their energies. Since the object in the box is a nonrelativistic free electron, its total energy would be entirely determined by its kinetic energy. Recall that the expressions for kinetic energy and momentum are

$$\text{kinetic energy} = \tfrac{1}{2}mv^2 = E$$

and

$$\text{momentum} = mv = p$$

leading to a relationship between energy and momentum, which is

$$E = \frac{p^2}{2m}$$

The energy of an nth state of motion E_n may then be expressed as:

$$E_n = \frac{p_n^{\,2}}{2m} = \frac{\left(\dfrac{h}{\lambda_n}\right)^2}{2m}$$

or

$$E_n = n^2\left(\frac{h^2}{8mL^2}\right) = n^2 E_0$$

An astonishing result of the preceding analysis is that the lowest energy that the electron may assume is not zero, implying that an electron confined to a finite region of space can never be in a state of rest. This result stands in sharp contrast to the particle-view analysis of the situation, which does admit a state of motionlessness as a natural state.

A nonzero lowest-energy state is also predicted by the uncertainty principle. If it were to attain a state of zero energy, the electron in the box would also have to attain a zero momentum, i.e., $p = 0$. But the uncertainty principle precludes establishing the momentum of the electron to be precisely zero. For such would imply that the position of the electron would be completely indeterminate, contradicting the initial assumption that the position of the electron is known within the length of the box:

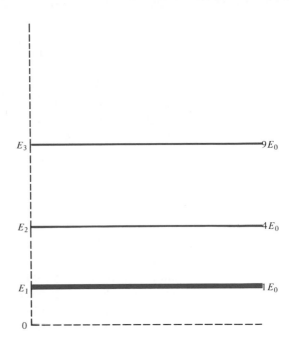

E_3 ———————————————9E_0

E_2 ———————————————4E_0

E_1 ———————————————1E_0

0

FIGURE 22–3. An electron, viewed as a wave, can exist only in a discrete set of quantized energy states.

that is, $\Delta x = L$. The electron in the box must therefore be viewed as having a nonzero momentum:

$$\Delta p = \frac{h}{L} \neq 0$$

As its momentum can never be observed to be zero, the electron cannot exist in a state of zero energy. In general, no quantum system confined to a finite region of space can ever exist in a state of motionlessness.

The wave-view of the electron dictates not only that the electron in a linear box shall never exist in a state of zero energy, but also that its energy shall be quantized according to $E_n = n^2 E_0$, where $E_0 = h^2/8mL^2$ and n is an integer. In general, any quantum system, confined to a finite region of space, would assume a discrete set of energies. The quantization of energy is thus an inescapable consequence of the wave nature of matter.

The Schrödinger Wave Equation

Schrödinger was particularly impressed with de Broglie's hypothesis of matter waves. His excitement was perhaps heightened by his deep knowledge of wave phenomena in general, and by his facility with the mathematical methods dealing with them.

The Schrödinger Theory of the Atom

233

As a historical sideline, it is interesting to note that while Schrödinger pursued the abstract idea of matter waves for their possible applicability to the atom, Heisenberg proceeded along a different approach, guided much more by experimental data. Heisenberg's success, in 1925, consisted in constructing a theoretical scheme into which to fit the observed atomic spectra. Heisenberg's theory of the atom is called matrix mechanics and is completely equivalent in physical content to Schrödinger's wave mechanics, which was published in 1926.

In developing his theory, Schrödinger approached the problem of the atom in a more abstract, mathematical way. He was, in a sense, guided more by pure thought than by experiment. The first equation he achieved was a relativistic wave equation for matter waves. Upon applying it to the electron of the hydrogen atom, however, he found that the initial results did not agree with experiment.[1] So deep was his disappointment that he abandoned the work for some months. The event to rekindle his interest in the work was an invitation to give a colloquium on the de Broglie wave. In the course of reviewing the idea of matter waves Schrödinger modified his original wave equation so as to be nonrelativistic. When he applied this nonrelativistic wave equation to the hydrogen atom, the results were in complete agreement with experiment.

In the Schrödinger theory the wave associated with a physical entity is called a *wave function*. A wave function for a quantum system results when the Schrödinger equation is solved for the particular system. A wave function is, then, viewed as containing all the information which may be known about the associated system within the restrictions of the uncertainty principle.

The Wave Functions of a Confined Electron

Consider again an electron confined to a linear box with impenetrable walls. When the Schrödinger wave equation is applied to such an electron, the wave functions describe standing waves, similar to those on a string with fixed ends. The condition imposed on the solution of the Schrödinger equation is that the wave function must vanish at the boundaries of the box, namely, at the walls. Such a "boundary condition" leads to a discrete

[1] The discrepancy between Schrödinger's relativistic wave equation and experiment was later resolved by taking into account the intrinsic spin of the electron.

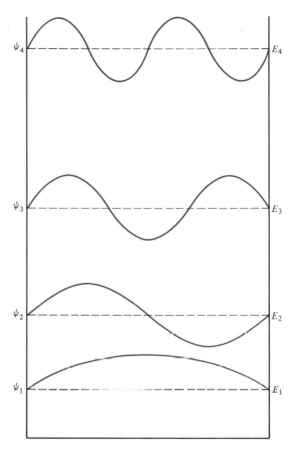

FIGURE 22–4. Stationary-state wave functions, ψ_n, for an electron confined in a linear box.

set of energies for the electron, and each standing wave characterized by a definite energy is said to describe a *stationary* state. In the Schrödinger theory, then, an electron confined to a linear box is described by an infinite set of stationary-state wave functions.

The Probabilistic Interpretation of the Wave Function

Strange as it may seem, the Schrödinger wave function does not lend itself to a direct physical interpretation. Rather, the wave function is a purely mathematical function which simply satisfies the Schrödinger equation in a given situation.

The Schrödinger Theory of the Atom

The physical interpretation of the wave function arises from the square of the wave function. The square of the wave function describes the probability of finding an electron, or any quantum object, at a specific point in space at a given time.

Take, for instance, a stationary-state wave function, Ψ_3, in our example of a confined electron. $(\Psi_3)^2$ then gives the probability density, or the probability per unit length, of finding the electron at any point in the region between the walls. Notice that the electron in the $n = 3$ state of motion would never be found at the walls, or at two other points in between. On the other hand, the electron has the greatest probability of being found at the center as well as at two other positions. If it exists in the $n = 2$ state of motion, however, the electron would have a zero probability of being found at the center.

The probabilistic interpretation of the wave function is fundamental to the Schrödinger theory. First proposed by Max Born in 1926, this particular approach has helped make the Schrödinger theory physically transparent.

The Hydrogen Atom

The hydrogen atom was the first crucial test for the Schrödinger wave equation. The problem consisted in describing the hydrogen electron as a de Broglie wave, taking into account the fact that the electron is confined to the vicinity of the nucleus by an electric force. It is the presence of the electric force that makes an atomic electron behave differently from an "electron in the box."

The matter of obtaining a Schrödinger wave function to describe an electron is essentially a mathematical problem. But mathematics alone is not sufficient to yield useful physical information. Conditions must, therefore, be imposed on the mathematical solutions so that they would be consistent with physical reality.

A condition to be imposed on the wave function is that the wave function must have only a single value at every point in space. That this must be so is almost obvious, for only then the probability of finding a particle would likewise have a single value at every point in space, and measurements would produce definite values. Another physical require-

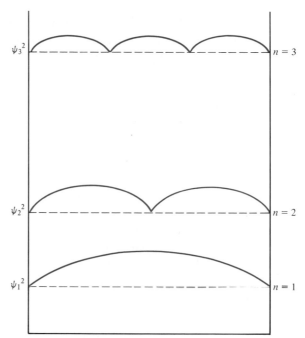

$\psi_3{}^2$ $n = 3$

$\psi_2{}^2$ $n = 2$

FIGURE 22–5. Probability density functions for an electron confined in a linear box.

$\psi_1{}^2$ $n = 1$

ment is that the wave function must be finite everywhere in space.[2] An infinite wave function would imply an infinite probability density, and would therefore be physically untenable.

The Schrödinger wave functions subject to these physical requirements describe a physically realizable situation. For an electron bound to the hydrogen nucleus, the Schrödinger theory yields a series of stationary-state wave functions, with each stationary state specified by a quantized energy. The energy associated with each stationary state of the hydrogen atom is given by a simple expression:

$$E_n = \frac{-E_0}{n^2}, \quad n = 1, 2, 3, \ldots$$

where E_0 is a constant characteristic of the hydrogen atom, and has a numerical value of 13.6 eV. Comparing this expression with Bohr's equation of energy quantization, we observe that the energies predicted by the new theory are identical with those predicted by the Bohr model. This result was the first successful test of the Schrödinger wave equation.

[2] A wave function could possibly be infinite at one point in space provided the total probability in a finite region remains finite.

The Schrödinger Theory of the Atom

The description of the motion of the hydrogen electron requires three coordinates, because it exists physically in a three-dimensional space. The stationary-state wave functions are likewise specified by a set of three numbers. Called quantum numbers, they are designated by n, l, and m, and assume integral values subject to the following conditions:

$$n = 1, 2, 3, 4, \ldots$$
$$l = 0, 1, 2, 3, \ldots, (n-1)$$
$$m = -l, -(l-1), \ldots, 0, 1, \ldots, l$$

These restrictions on the quantum numbers emerge in a natural way when the wave functions are required to conform to physical reality.

It is interesting to note that although the stationary state Ψ_{nlm} is specified by the values of all three quantum numbers, the energy of the state depends only on the value of n. For example, the stationary state $\Psi_{2,0,0}$ has the same energy as the stationary state $\Psi_{2,1,1}$. But either state, of course, has a higher energy than $\Psi_{1,0,0}$.

The Radial Probability Densities of the Hydrogen Atom

An important feature of the probability density of a stationary state is that the probability density does not depend on the time. The square of the stationary-state wave function therefore yields a static picture of the electron's state of motion.

Now the static picture of the hydrogen atom depends on all three quantum numbers, because the wave function itself does. However, if, for the sake of simplicity, the quantum number m is suppressed, the result gives the probability density as a function of the radial distance from the nucleus. Such information would clearly be very enlightening. We, therefore, discuss the radial probability densities of several hydrogen stationary-state wave functions.

From Figure 22-6 we notice that the radial probability densities are reasonably concentrated in finite regions. The electron is, in a sense, smeared in the atom. As the spatial distributions of the electron have the appearance of a cloud, they are often referred to as electron clouds.

Let us now ask if the probability densities of the Schrödinger theory are in any way related to the electron orbits of the Bohr theory. Now the

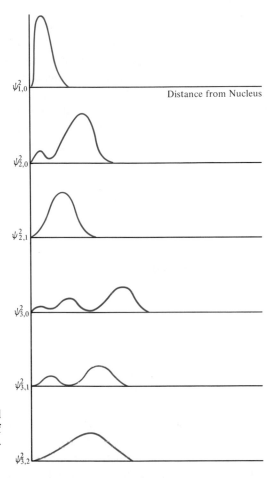

$\psi_{1,0}^2$

Distance from Nucleus

$\psi_{2,0}^2$

$\psi_{2,1}^2$

$\psi_{3,0}^2$

$\psi_{3,1}^2$

FIGURE 22–6. The radial probability functions of the electron in the hydrogen atom.

$\psi_{3,2}^2$

first stationary state, $\Psi_{1,0}$, gives rise to an appreciable electron cloud at a certain distance away from the nucleus, namely, $r = 0.53 \times 10^{-8}$ cm. Recalling that this number is exactly the radius of the first ($n = 1$) circular orbit of the Bohr model, it is certainly not unreasonable to associate this fuzzy electron cloud with the first Bohr orbit. The second stationary state, $\Psi_{2,0}$, has two regions of appreciable electron cloud. The corresponding Bohr orbit is an ellipse. The two peaks may be interpreted in terms of de Broglie's idea and arise because the distance along the elliptical path is approximately equal to twice the de Broglie wavelength. The $\Psi_{2,1}$ state corresponds to the $n = 2$ circular orbit. One can now appreciate why the Bohr model was initially successful, but its inadequacy is also transparent in the light of the Schrödinger theory.

The Schrödinger Theory of the Atom

The stationary states also give the charge distributions of the atom. The distribution of charge in an atom is determined by the probability density of the electron. Since the probability density of a stationary state does not change with time, a stationary state describes a static charge distribution. A static charge distribution, however, does not radiate; only a dynamic charge distribution radiates. An atom existing in a stationary state, therefore, cannot radiate photons. Bohr's ad hoc assumption that an electron orbiting in an allowed orbit does not radiate (recall that it was this hypothesis that came in violent conflict with Maxwell's electromagnetic theory) is thus totally unnecessary in the Schrödinger theory of the atom.

Nonstationary States

A unique feature of wave motion is that waves superpose to produce an interference effect. The stationary-state wave functions of the Schrödinger theory may also superpose to yield a new wave function. Of fundamental importance is the fact that this new wave function must also describe an observable state.

To begin with, let us construct a new state, Ψ_{1+2}, out of the superposition of two stationary states, Ψ_1 and Ψ_2, with energies E_1 and E_2, respectively. A physical interpretation of the new state is achieved when its probability density is calculated. The new probability density, unlike that of the stationary state, is found to vary with time. Its time dependence may be argued as follows. Since we have constructed the new state as the sum of two stationary states, namely,

$$\Psi_{1+2} = \Psi_1 + \Psi_2$$

the square of the new wave function would be

$$\Psi^2_{1+2} = \Psi_1{}^2 + \Psi_2{}^2 + 2\Psi_1\Psi_2$$

Now we know that Ψ_1 and Ψ_2 are time dependent, although $\Psi_1{}^2$ and $\Psi_2{}^2$ are not. Thus the time dependence of Ψ^2_{1+2} arises from the last term, the product of Ψ_1 and Ψ_2.

It turns out that the time dependence of the superposed probability density is oscillatory. The frequency of oscillation is determined by the energies of the stationary states being superposed, and is given by

$$f_{2,1} = \frac{E_2 - E_1}{h}$$

where h, as usual, denotes Planck's constant. The superposed wave function Ψ_{1+2} may then be interpreted as describing an electron that *oscillates* between two stationary states having energies E_2 and E_1 with a frequency $f_{2,1}$. Such an electron is said to exist in a *nonstationary* state.

Absorption of Radiation

The question now arises: How is the superposition of stationary-state wave functions achieved in an atom? As a suggestive analogy, recall a child on the swing. The child's ability to swing requires the presence of an adult with a sense of rhythm. Once the swing is set in oscillatory motion, the child must be given a push in harmony with the natural frequency of the swing. Otherwise the child would cease to swing.

In the case of the atom the external agent to set an electron in oscillatory motion is electromagnetic radiation. An atom possesses a discrete set of natural, or resonant, frequencies determined by the difference in energy between various stationary states. If an incident electromagnetic wave has a frequency equal to one of the natural frequencies of the atom, the incident radiation links the appropriate stationary states, thus superposing the corresponding wave functions. The resulting superposed state is a nonstationary state and describes an electron which oscillates between the linked stationary states. A transfer of energy is thus achieved from the electromagnetic radiation to the atom. Such a process is referred to as the resonant absorption of radiation by an atom.

Returning to the hydrogen atom again, let us suppose that the atom is initially in the Ψ_2 state. If it is exposed to red light, for example, the light would superpose the Ψ_2 and Ψ_3 states, because the natural frequency of oscillation between these two states is equal to the frequency of red light. The electron would thus be set in oscillatory motion. As the red light transfers its energy to the hydrogen atom, the electron would finally end up in the higher energy state Ψ_3. Thus, the absorption of red light by a hydrogen atom—and the absorption of radiation by atoms in general—is beautifully described by the Schrödinger theory.

Induced Emission of Radiation

When red light falls on a hydrogen atom in the Ψ_2 state, the hydrogen atom absorbs the radiation and is excited to the higher energy state, Ψ_3. But suppose now that red light falls on a hydrogen atom which is initially in the Ψ_3 state. What would happen to the electron then? Red light cannot be absorbed, because none of the natural frequencies corresponding to the superposition of Ψ_3 and a still higher stationary state is equal to the frequency of red light.

Red light, however, can superpose Ψ_3 and Ψ_2, thus causing the electron to oscillate between these two states. The electron would, then, exist in an oscillatory, nonstationary state until a transition to the lower energy state is achieved. A downward transition, however, must be accompanied by the emission of a photon—in this case, red light. Such a process is referred to as the *induced,* or stimulated, emission of radiation, as it indeed is induced by the presence of radiation from an external source.

As far as the atom is concerned, there is no difference in the basic mechanism of absorption and induced emission of radiation. The only incidental difference is the initial state of the atom when it is subjected to the perturbing influence of external radiation. An atom undergoes absorption if it happens to be in the lower energy state of the superposed states, and induced emission if it is initially in the upper energy state.

Spontaneous Emission of Radiation

Spontaneous emission of radiation is an inherent property of the atom. Contrary to induced emission, spontaneous emission does not require the perturbing influence of an external radiation.

The emission of radiation from an atom arises from the oscillatory motion of an electron. Take, for example, a hydrogen electron in the Ψ_3 state. If it oscillates between Ψ_3 and Ψ_2, the electron would constitute an oscillating electric charge and would therefore radiate energy. The oscillatory motion ceases when the energy of the emitted radiation is built up to a magnitude of $hf_{3,2}$; that is, when a photon with an energy equal to $(E_3 - E_2)$ is produced.

The question then arises: How long does the oscillatory behavior of an atomic electron persist? The duration of oscillatory motion defines the lifetime of the excited state in which an electron exists initially. The lifetime of an excited state depends, in general, on the lower energy state to which the electron makes a transition. The excited states, leading to the emission of visible light, have lifetimes in the range of 10^{-7} seconds to 10^{-9} seconds. The typical lifetime of an excited atom is about 10^{-8} seconds.

The concept of the lifetime of an excited state is important, because it is related to the probability of an atom emitting a particular photon. The probability of emission, in turn, determines the number of particular photons in relation to the total number of excited atoms, and, therefore, yields the intensity of the emitted radiation. The intensities of the spectral lines are thus intimately related to the lifetimes of excited states.

To be specific, let us examine the lifetime of the $n = 3$ excited state of the hydrogen atom. The electron initially in the $n = 3$ state may make a downward transition to the $n = 2$ state, or to the $n = 1$ state. The lifetime of the $n = 3$ state is shorter for a transition to the $n = 2$ state than for a transition to the $n = 1$ state. During a given interval of time, then, a transition to the $n = 2$ state is more likely to occur than to the $n = 1$ state. Consequently, if there are many hydrogen atoms in the $n = 3$ state, there would be emitted more photons having an energy equal to $(E_3 - E_2)$ than those with an energy of $(E_3 - E_1)$. Thus, the Schrödinger theory, in predicting the seemingly abstract lifetimes of excited states, explains why one spectral line is observed to be more intense than another spectral line. By contrast, the Bohr model of the atom is totally silent on this point.

We have thus witnessed a fundamental theory in action. We have gained a thrilling insight into the nature of the atom. The picture the Schrödinger theory unveils for us is a beautful one. As a coherent picture of the quantum world, the Schrödinger theory is a towering monument to the creative vision of man.

QUESTIONS

22–1. Consider an electron confined to a linear box with impenetrable walls.

(a) If the length of the box is increased, how would the energy of the lowest state change?

Answer: decrease

The Schrödinger Theory of the Atom——————————————————

(b) Under what conditions would the ground state energy of the electron approach zero?

Answer: when the length of the box is made large

22–2. Consider a marble in a match box.

(a) What is its lowest state of motion?

Answer: state of rest

(b) Why is the marble able to achieve a state of rest, whereas a confined electron cannot?

Answer: mass of the marble is so much larger than electron mass, and Planck's constant is a small number

22–3. Give a physical interpretation to the first three wave functions describing an electron confined to a linear box.

22–4. Can all the allowed orbits of the Bohr atom be regarded as stationary states? Explain.

Answer: yes

22–5. Explain why an electron in a stationary state does not radiate.

22–6. (a) Explain how a nonstationary state is constructed.

Answer: by superposing stationary states

(b) Explain why a nonstationary state describes an oscillating charge distribution.

22–7. Describe the mechanism of resonant absorption and induced emission of radiation.

22–8. What is the difference between induced and spontaneous emission of radiation?

22–9. Suppose you are observing 1000 excited hydrogen atoms, all initially in the $n = 3$ state. Suppose further that all downward transitions are equally probable. How many photons would you observe when all atoms have returned to the ground state?

Answer: 1500 photons

22–10. Explain the meaning of the intensity of a spectral line in terms of photons?

22–11. List the weaknesses of the Bohr theory. How are they removed or resolved by the Schrödinger theory?

22–12. The Schrödinger theory is an example of a fundamental physical theory. Comment on the qualities of a fundamental theory.

SUGGESTIONS FOR FURTHER READING

COOPER, L. N. *An Introduction to the Meaning and Structure of Physics,* short edition. New York: Harper & Row, Publishers, 1970. (Chapters, 33, 34, and 35.)

DIRAC, P. A. M. "The Evolution of the Physicists' Picture of Nature," *Scientific American* (May 1963).

HEISENBERG, W. *Physics and Beyond.* New York: Harper & Row, Publishers, 1971.

SCHAWLOW, A. L. "Optical Masers," *Scientific American* (June 1961).

SCHRÖDINGER, E. "What Is Matter?" *Scientific American* (September 1953).

23

The Atomic Nucleus

In the study of the structure of the atom, the nucleus is viewed as a structureless physical entity. To suggest that the nucleus also has an internal structure, however, would constitute no great surprise. In fact, a wealth of information exists about the nucleus, and it may be claimed that the structure of the nucleus is fairly well understood.

Nuclear Size

As a constituent of the atom, the nucleus may indeed be considered as a particle of infinitesimal size. Compared to the average atomic size of 10^{-8} cm, the nucleus has a radius of the order of 10^{-13} cm, this being some 100,000 times smaller in size than the atom. The size of a nucleus, however, varies from one nucleus to another, contrary to the roughly identical size for all atoms. It has been established experimentally that the radius of a nucleus is given by

$$r = r_0 A^{1/3}, r_0 = 1.2 \times 10^{-13} \text{ cm}$$

where A denotes the number of the total constituents of the nucleus. A nucleus containing a higher number of constituent particles would then have a correspondingly larger size.

Constituents

The fundamental constituents of the nucleus are protons and neutrons. The proton carries one unit of positive charge, whereas the neutron carries zero charge. Aside from the difference in charge, the proton and the neutron are quite similar. Inside the nucleus, in fact, they essentially lose their individual identity, so that they are referred to collectively as nucleons. They may thus be regarded as two different charge states of the nucleon.

The total number of nucleons in a nucleus determines the mass of the nucleus and is thus called the mass number of the nucleus, designated by the symbol A. The number of protons, on the other hand, determines the total charge of a nucleus and, therefore, also determines the number of electrons in the corresponding atom. This number is, then, nothing more than the atomic number, usually denoted by Z. A nucleus is specified when its mass number A and atomic number Z are given. An oxygen nucleus, for example, which consists of eight protons and eight neutrons, is symbolically represented by $^{16}_{8}O$.

The mass of an atom is concentrated in the nucleus. The contribution of the electrons to the total mass of an atom is negligible, because the mass of the nucleon is about 1,800 times as great as the mass of the electron. It is, in fact, the large magnitude of nucleon mass that enables the nucleons to be bound in a small region of space, namely, the nucleus. If a nucleon were a bit less massive, the nuclear force would not be strong enough to bind nuclei.

The Yukawa Theory of Nuclear Force

The concept of force constitutes an integral part of any description of motion. Newton's laws of motion, for example, could not describe the motions of the planets apart from the specification of gravitational force. Similarly, the Schrödinger wave equation yields a correct description of the motions of atomic electrons, because the electric force which the nucleus exerts on the electrons is known independently of the Schrödinger theory. It is, then, not unreasonable to expect that a correct description of the motion of nucleons inside the nucleus would also require a knowledge of the force that holds the nucleus together.

Nuclear force has two distinctive features. It is, first of all, a force of great strength. Secondly, it is a short-range force, being effective only for very short distances. Nuclear force does not extend much beyond 10^{-13} cm from its source, whereas both electromagnetic and gravitational forces extend out to infinity.

The first successful attempt at arriving at a quantum description of the nuclear force was made in 1935 by the Japanese theoretical physicist, H. Yukawa. He attacked the problem essentially in a manner similar to that used in the well-established quantum description of the electromagnetic force.

The quantum nature of the electromagnetic force is described by a theory called quantum electrodynamics. Quantum electrodynamics was initially proposed by Dirac in 1926 to explain the absorption, emission, and scattering of radiation by atoms. In the Schrödinger theory the force between the electron and the nucleus is introduced in an ad hoc fashion, because the electromagnetic force is not subjected to any sort of quantization. In a sense, the Schrödinger theory thereby suffers a certain lack of consistency. According to quantum electrodynamics, on the other hand, the electromagnetic interaction between particles, say between an electron and a proton, arises indirectly: the electron described by a de Broglie wave interacts with the quantized electromagnetic field, and this field in turn interacts with the proton, which is also described by a de Broglie wave. A quantized electromagnetic field is exactly what is meant by a photon. In quantum electrodynamics, the electromagnetic interaction between two charged particles is thus viewed as arising via an exchange of photons.

The idea that the electromagnetic interaction between charged particles is mediated by the exchange of photons was adapted by Yukawa to the short-range nuclear interaction between nucleons. He proposed that a field is also associated with the nucleon and that the interaction between nucleons arises through the exchange of a quantum of this field. An exchange occurs when one nucleon emits a nuclear quantum and another absorbs it. Now the long-range nature of the electromagnetic interaction is viewed, in quantum electrodynamics, as a result of the zero rest mass of the photons which mediate between charged particles. Pursuing the analogy of quantum electrodynamics further, Yukawa then proposed

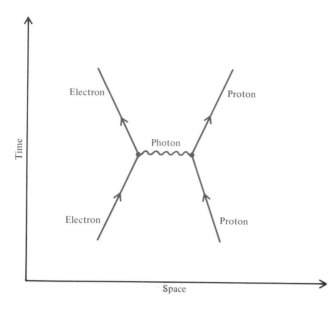

FIGURE 23–1. The world lines for an electron–proton interaction via the exchange of a virtual photon.

that if nuclear force is to have a short-range character, the quantum of a nuclear field must have a finite mass.

Yukawa's model envisions one nucleon emitting a nuclear quantum of finite mass to be later absorbed by another nucleon. Such an exchange, however, violates the conservation of energy, because the nuclear quantum must carry energy. The energy, ΔE, carried by the quantum must at least be equal to its rest energy of the quantum,

$$\Delta E = m_\pi c^2$$

where m_π denotes the rest mass of the exchanged quantum. One immediate reaction is that such an interaction could not describe a physical situation, because no physical process can violate the principle of energy conservation. Such a reaction is well justified in the realm of macroscopic physics, but the quantum world is governed by the uncertainty principle. The uncertainty principle shows that violation of energy conservation by an amount ΔE may be permitted if such violation lasts no longer than a certain length of time Δt, limited by the uncertainty relation

$$\Delta E \Delta t = \hbar$$

where $\hbar$ is a compact notation for $h/2\pi$. A particle which owes its transitory existence to the uncertainty principle is called a virtual particle.

The rest energy of the virtual nuclear quantum can now be estimated

Physics: The Fabric of Reality

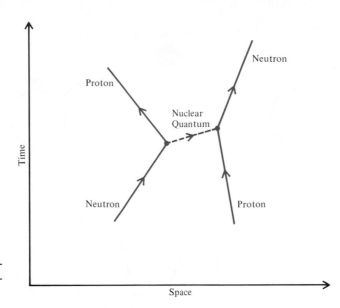

FIGURE 23–2. Exchange of a nuclear quantum between a neutron and a proton.

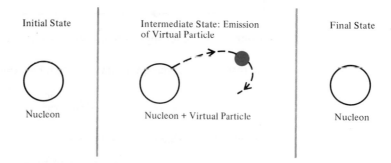

Initial State

Intermediate State: Emission of Virtual Particle

Final State

Nucleon

Nucleon + Virtual Particle

Nucleon

FIGURE 23–3. The virtual activity of a nucleon.

with the help of the uncertainty principle. Because no particle, virtual or real, can travel faster than the speed of light, the virtual nuclear quantum can traverse a distance no farther than that given by

$$R = c\Delta t$$

Two nucleons exchanging such a quantum must then be within a distance R. Put another way, a nuclear interaction must take place within this distance. But this distance is known experimentally to be about 1.4×10^{-13} cm, or

$$R = 1.4 \times 10^{-13} \text{ cm}$$

The Atomic Nucleus

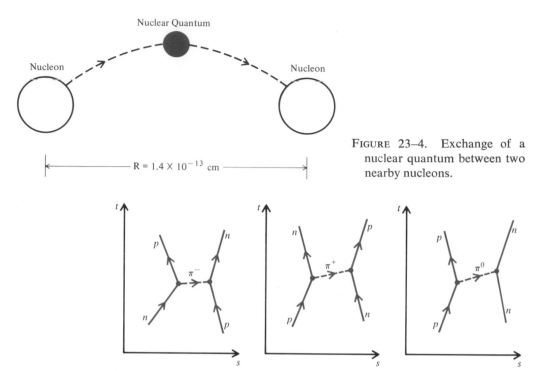

FIGURE 23–4. Exchange of a nuclear quantum between two nearby nucleons.

FIGURE 23–5. Exchange of pions between nucleons.

The rest energy of the quantum may then be expressed as

$$\Delta E = \frac{\hbar c}{R}$$

The numerical value comes out to be

$$m_\pi c^2 = 140 \text{ MeV}$$

In terms of the electron mass, this new particle has a mass about 275 times as great as the electron. Recalling that the nucleon has a mass about 1,800 times the electron mass, the new particle's mass falls somewhere in the middle between the electron and the nucleon. It was therefore named a meson, after a Greek word meaning "middle." The quantum field associated with the meson is aptly called a meson field, and the Yukawa theory of the nuclear force is referred to as a meson theory. The meson theory views the nuclear interaction between nucleons as arising through the mediation of a meson field, or equivalently through the exchange of virtual pions, as the Yukawa quanta are now called.

There are three kinds of pions, having a different electric charge: positive, negative, or neutral. The emission of a positively charged pion, π^+, by a proton would change it to a neutron, and a neutron, absorbing it, would turn into a proton. A neutron, on the other hand, could emit a negatively charged pion, π^-, thus turning into a proton. Neither the proton nor the neutron would suffer a change in its charge state if a neutral pion, π^0, is exchanged.

Charge Structure of the Nucleon

The meson theory views the nucleon as existing in an constant state of activity. The nucleon emits a virtual pion and reabsorbs it within the time limit governed by the uncertainty principle. It would then be natural to wonder if the proton and the neutron also have internal structures. As a first approximation, they may be pictured as consisting of a dense core, surrounded by a spinning pion cloud.

The existence of a pion cloud in the proton has been observed. The experiment consists of bombarding a target of gaseous hydrogen with a high-energy electron beam. The higher the energy of the probing electron, the shorter its de Broglie wave length is. But the shorter the wavelength of the probing wave, the finer would be the details of the nucleon structure revealed by it. The charge structure of the proton is unfolded from the diffraction pattern of the scattered electrons. The experimental result is that the proton has a dense positively charged core, surrounded by a positively charged pion cloud.

The neutron also has spinning pion clouds. These pion clouds give rise to a charge structure in the neutron. Deep inside the neutron the charge density is positive The charge density then turns negative, as one goes out farther from the center. At the edge of the neutron the charge density turns positive again. These different charge regions inside the neutron add so as to yield a net zero charge for the neutron.

Observation of Real Pions

Pions in the nucleon or in the nucleus are not directly observable. They owe their existence to the uncertainty principle, which allows the principle of energy conservation to be violated. If, however, the required

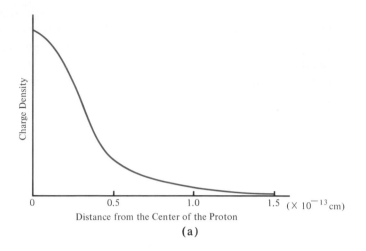

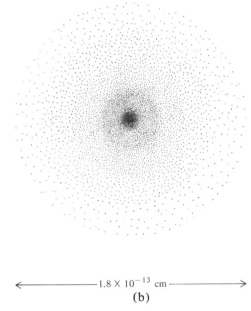

(b)

FIGURE 23–6. (a) The charge distribution within a proton and (b) a schematic sketch of the charge structure of a proton.

energy is supplied to a nucleon or a nucleus, pions may attain the status of a real particle, and thus become directly observable.

Charged pions were first observed by C. F. Powell and collaborators in 1947 in cosmic rays. Cosmic rays may be divided into two categories: the primary cosmic rays, which are predominantly protons, entering the earth's atmosphere from outer space, and the secondary cosmic rays, consisting of all kinds of particles produced in the atmosphere upon bombardment by the primary protons. One of the secondary particles is

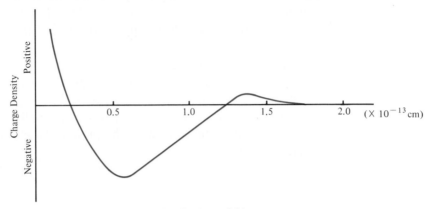

FIGURE 23–7.   The charge distribution within a neutron.

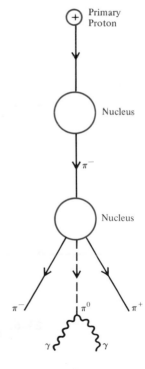

FIGURE 23–8.   Schematic representation of the production of pions in the atmosphere.

The Atomic Nucleus

the pion. Pions can also be produced in high-energy nucleon–nucleon collisions in accelerators. In fact, the identification of the neutral pion was achieved in 1950 in an accelerator experiment as well as in cosmic ray studies.

Positively charged pions have a rest mass 274 times as great as the mass of the electron. Negatively charged pions are the antiparticle counterparts of the positively charged pions. The mass of the neutral pion is somewhat lower, being approximately 264 times the mass of the electron.

Pions are radioactive particles. The charged pion has a half-life of 1.78×10^{-8} seconds, and the neutral pion has a half-life of 0.7×10^{-16} seconds. The radioactive character of the pion, however, is not an unusual feature, as all fundamental particles are radioactive with the exception of four particles, namely, the proton, the electron, the photon, and the neutrino.

QUESTIONS

23–1. The volume of a sphere of radius r is given by $\frac{3}{4}\pi r^3$. If the shape of the nucleus is assumed to be spherical, how would the nuclear volume vary with the nuclear mass number A? Explain.
Answer: Directly proportional to A

23–2. Explain why if a nucleon were a bit less massive, the nuclear force would not be strong enough to bind nuclei. (Hint: Apply the uncertainty principle.)

23–3. Comment on the role of reasoning by analogy in creative thought processes.

23–4. The graviton is a quantum of the gravitational force. What would the rest mass of the graviton be? Explain.

23–5. The Compton wavelength of the pion is defined to be $\hbar/m_\pi c$, where $\hbar = h/2\pi$, c is the speed of light, and m_π is the rest mass of the pion. What is its numerical value?
Answer: 1.4×10^{-13} cm

23–6. Pions can be created in a high-energy proton–proton collision. What nucleons would emerge as reaction products upon the creation of the positively charged pion, the negatively charged pion, and the neutral pion, respectively? (Consider only the simplest possibilities.)

Answer: p^+ and n^0 with π^+

23–7. (a) Explain why the pion must have a finite mass if it is to mediate a nuclear interaction.

(b) If the pion were a particle of zero rest mass, how would the nature of the nuclear interaction be changed?

SUGGESTIONS FOR FURTHER READING

BETHE, H. A. "What Holds the Nucleus Together?" *Scientific American* (September 1953).

FORD, K. W. *Basic Physics*. Waltham, Mass.: Blaisdell Publishing Company, 1968. (Chapter 25.)

HOFSTADTER, R. "The Atomic Nucleus," *Scientific American* (July 1956).

MARSHAK, R. E. "The Nuclear Force," *Scientific American* (March 1960).

24

Radioactivity

The chance discovery of radioactivity by the French physicist Henri Becquerel in 1896 marks the beginning of the nuclear era. The nucleus is a bundle of concentrated matter. Bound by the strongest force in nature, the nucleus thus provides the greatest source of energy known to man.

Binding Energy

The concept of binding energy is very useful in the physics of the nucleus. Suppose we take a deuteron, the nucleus of the deuterium atom, which consists of a proton and a neutron, and break it up into constituent parts. It would be observed that the separation of a deuteron into its constituent proton and neutron requires a minimum energy of 2.2 MeV. This energy is referred to as the binding energy of the deuteron. The breaking-up process involves a transition from the initial state consisting of a deuteron and a 2.2 MeV photon to the final state consisting of a proton and a neutron as independent particles.

$$\text{photon} + \text{deuteron} \rightarrow \text{proton} + \text{neutron}$$

The conservation of energy requires that the total energy of the initial state remain the same as that of the final state. This implies that the sum of the photon energy and the rest energy of the deuteron must equal the sum of the rest energies of the proton and the neutron.

$$E_{\text{ph}} + M_d c^2 = M_p c^2 + M_n c^2$$

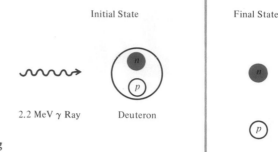

Initial State | Final State

2.2 MeV γ Ray   Deuteron

FIGURE 24–1. Binding
energy of the deuteron.

where M denotes the mass of a particle and c is, as usual, the speed of light. The sum of the masses of the separated particles is then greater than the mass of the composite nucleus. The extra mass arises from the conversion of the energy of the photon.

Conversely, if a proton and a neutron are fused to form a deuteron, an energy of 2.2 MeV is released. This energy is a result of the conversion of the excess mass which does not get accommodated in the mass of the deuteron.

$$\text{proton} + \text{neutron} \rightarrow \text{deuteron} + \text{photon}$$

The source of the binding energy is the conversion of mass. The total binding energy of an oxygen nucleus, for example, is computed in the following way.

$$E_{\text{binding}} = 8M_p c^2 + 8M_n c^2 - Mc^2$$

where M denotes the mass of an oxygen nucleus, $^{16}_{8}\text{O}$. The calculation yields a binding energy of 128 MeV for the oxygen nucleus. An energy of 128 MeV is thus required to break up an oxygen nucleus into its sixteen constituent nucleons, or conversely, a similar energy would be released upon fusion of eight protons and eight neutrons into an oxygen nucleus. Perhaps more useful is the concept of the binding energy per nucleon, which is determined by dividing the total binding energy by the number of nucleons. The binding energy per nucleon for oxygen is about 8 MeV.

It is a remarkable fact that the binding energy per nucleon is about the same for all stable nuclei. Marked exceptions occur in very light nuclei, and their considerably lower values imply that the fusion of light nuclei into heavier nuclei would result in the release of considerable energy.

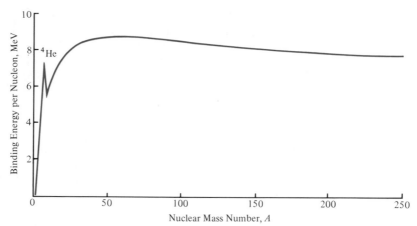

FIGURE 24-2. The binding energy per nucleon versus the nuclear mass number.

Beyond the mass number, $A = 50$, the average binding energy per nucleon decreases slightly as the mass number increases. A very heavy nucleus, splitting into two nuclei of intermediate mass, would thus be accompanied by a release of energy.

Nuclear Energy States

The nucleus is a complex physical structure. Nuclear structure, like atomic structure, is understood in terms of the concept of energy levels. Each nucleus is envisioned as having its own set of quantized energy levels. Nuclear energy levels are typically separated by millions of electron volts, in contrast to atomic energy level spacings of a few electron volts. This fact, of course, comes as no surprise, because nuclear force, as the strongest force known to exist in nature, is much greater in strength than the electromagnetic force that holds the atom together.

The decay mode of the nucleus is more varied than that of the atom. An excited atom decays by emitting a photon. The nucleus, on the other hand, has three modes of decay. Some nuclei decay by emission of particles such as the alpha and beta particles, whereas all nuclei may decay by emission of photons.

It is customary to measure the rate of nuclear decay in terms of the concept of the half-life. The half-life of a radioactive nucleus is defined

Radioactivity

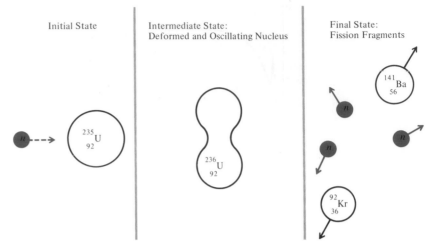

FIGURE 24–3. When uranium-235 captures a neutron, a compound nucleus, U-236, is formed, which deforms and oscillates, subsequently fissioning into nuclei of intermediate size.

as the length of time elapsed during the decay of one half of the initial number of nuclei. The half-life of a radioisotope is not affected by any external influences. The half-life of carbon-14, for example, is about 5,770 years, whether such carbon nuclei are found in the upper atmosphere or in underground coal.

Gamma Decay

The average binding energy per nucleon for oxygen-16 is about 8 MeV. This, however, does not imply that an energy of 8 MeV would be required to separate the least tightly bound nucleon from the nucleus. On the contrary, observation indicates that, on the one hand, an energy of 12.1 MeV is required to separate the least tightly bound proton, and on the other hand, an energy of 15.6 MeV is required to release the outermost neutron from the oxygen-16 nucleus. These values of separation energy suggest that neutrons are bound more tightly than the protons to the nucleus, a reason for which may be traced to the electrical repulsion among protons.

Suppose now that a nucleus is excited by an energy that is less than the separation energy of the least tightly bound nucleon. The nucleus then would not suffer the loss of a nucleon. Such a nucleus, however, would be excited to a higher energy state internally and would return to its normal state by emitting photons. These photons are typically more energetic than X-rays and are called gamma rays. The process of de-excitation by gamma emission is aptly referred to as gamma decay.

Gamma transitions in a nucleus are analogous to the transitions that occur between the quantized energy states of an atom. The energy of the gamma ray is determined by the energy states linked by a transition,

$$hf = E_{\text{upper}} - E_{\text{lower}}$$

In constructing a nuclear energy level diagram it is customary to assign zero to the ground state of the nucleus, and positive values of energy to all excited states.

Alpha Decay

Nuclear stability is achieved by an attractive nuclear force that holds the nucleons together in the nucleus. Among the protons, however, there operates an additional force of electrical repulsion. In the nuclei of inter-mediate size the nuclear force dominates over the electrical repulsion. Heavy nuclei, on the other hand, tend to become unstable, as the inc-creasing number of protons increases electrical repulsion. Furthermore, the effect of nuclear size on the attractive nuclear force is to render it less effective, because nuclear force is strong only over a short range.

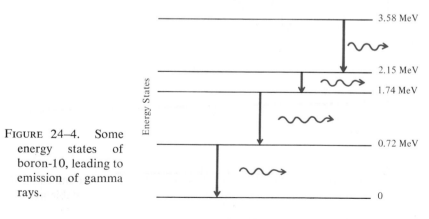

FIGURE 24–4. Some energy states of boron-10, leading to emission of gamma rays.

In fact, the highest number of protons that can be accommodated in a stable nucleus is 83, which is the nucleus of bismuth. All nuclei heavier than bismuth are unstable.

Heavy nuclei have a larger number of neutrons than protons. Neutrons, suffering no electrical repulsion, help overcome the repulsion among the protons by providing only an attractive nuclear force. It thus seems that one way of achieving greater stability by a heavy nucleus would be by reducing the relative number of protons. Such a reduction is, in fact, accomplished by the emission of alpha particles. An alpha particle consists of two protons and two neutrons and is identical to the nucleus of the helium atom.

When an unstable nucleus decays into a different nucleus by emission of a particle, the former and the latter are, respectively, referred to as the

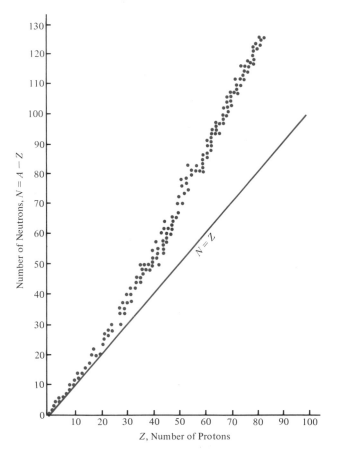

FIGURE 24–5. Chart of the known stable nuclei.

parent and the daughter nuclei. Alpha decay may then be represented by

$$\ce{^A_Z P} \rightarrow \ce{^{A-4}_{Z-2} D} + \ce{^4_2 He}$$

where P and D denote the parent and daughter nuclei. The preceding representation displays two conserved quantities explicitly. The subscripts, denoting the number of protons in each nucleus, are balanced, thus conserving electric charge: The total charge before decay is the same as the total charge after decay. The superscripts, on the other hand, denote the number of nucleons in the nucleus. Alpha decay then conserves the number of nucleons. In fact, all physical processes conserve the number of nucleons, thus giving rise to the principle of conservation of the nucleon number.

Alpha decay is accompanied by a release of energy. This energy appears in the form of the kinetic energy of the alpha particle and the daughter nucleus and is shared by them in a well-defined way, leading to the result that the emitted alpha particles are monoenergetic. A particular group of alpha particles in the decay of uranium-238, for example, all carry a kinetic energy of 4.18 MeV.

Beta Decay

Certain unstable nuclei achieve stability by ejecting an electron or a positron, or by capturing an atomic electron. Electrons and positrons are collectively referred to as beta particles. Boron and nitrogen are examples of beta emitters. They are particularly interesting, because they

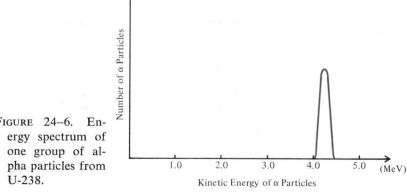

FIGURE 24–6. Energy spectrum of one group of alpha particles from U-238.

both transform into a common daughter nucleus of carbon. Their beta decay becomes transparent when the neutron and proton occupation levels of boron, carbon, and nitrogen are examined.

There are separate energy states available to the protons and neutrons in the nucleus. Each state can contain no more than two protons or two neutrons, because both protons and neutrons obey the Pauli exclusion principle. The accommodation of two particles in each energy level comes about by virtue of two possible spin states for a nucleon. Both the proton and the neutron have intrinsic spin, which assumes values $+\frac{1}{2}$ or $-\frac{1}{2}$ in units of $h/2\pi$.

Carbon-12 is a stable nucleus, because all the lowest available energy states are completely filled. Boron-12, on the other hand, has too many neutrons and is thus unstable. A more stable state would obtain if the single neutron in the highest energy state fills the single unoccupied proton state. But, in order to do so, the neutron must transform into a proton. It is observed that such a transition is accompanied by emission of an electron.

$$n^0 \rightarrow p^+ + e^-$$

Thus boron-12 achieves stability in the form of carbon-12 by emitting an electron. Similarly, nitrogen-12 has too many protons to be stable. Stability is achieved if the proton in the highest energy state makes a transition to a lower neutron energy state, thus becoming a neutron. Such a

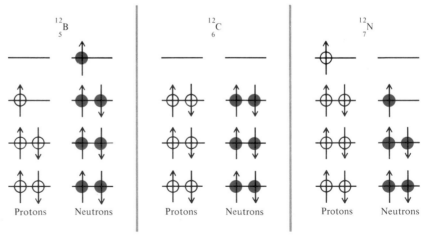

FIGURE 24–7. A schematic representation of the ground states of $^{12}_{5}\text{B}$, $^{12}_{6}\text{C}$, and $^{12}_{7}\text{N}$.

transformation, however, is observed to be accompanied by the emission of a positron.

$$p^+ \rightarrow n^0 + e^+$$

or

$$^{12}_{7}\text{N} \rightarrow {}^{12}_{6}\text{C} + {}_{+1}e$$

As in the case of alpha decay, energy is also released in beta decay. It seems, at first, that because the energy would be shared by the daughter nucleus and the beta particle, the beta particles would be monoenergetic. Furthermore, because the beta particles are so much lighter than the daughter nuclei, the recoils of the latter would be negligible, and essentially the entire energy released in beta decay would be carried by the beta particles.

Experimental measurements of the electron energy in beta decay indicate that the emitted electrons are not monoenergetic. In the case of boron-12, for example, the electrons are found to carry an energy anywhere between zero and a maximum 13.4 MeV. If beta decay were similar in form to alpha decay, the emitted electrons would all carry 13.4 MeV of energy, because such would be required by the conservation of energy. But, in fact, a great number of them carry only 10.0 MeV of energy. One must then ask what has happened to the missing energy of 3.4 MeV. It is certainly not carried by the recoiling nucleus, nor is it detectable with any apparatus. One is thus confronted with a situation where starting with a certain amount of energy, one ends up with a less amount of energy, with the difference in no way accounted for. Such a situation is extremely serious, indeed, as it violates nothing less than the principle of energy conservation itself.

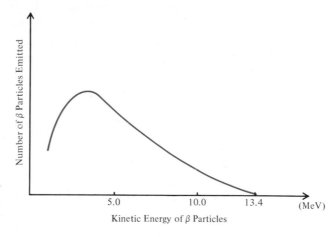

FIGURE 24–8. Continuous energy spectrum of electrons emitted in the beta decay of boron-12.

Radioactivity

Beta decay, viewed as a process of simple beta emission, violates two other well-established conservation laws: angular momentum (or spin) conservation and linear momentum conservation. In boron-12, for example, a neutron is transformed to a proton and an electron, but all the particles involved are particles with a spin $\frac{1}{2}$. Now each particle can be spinning up ($+\frac{1}{2}$) or spinning down ($-\frac{1}{2}$). No combination of proton and electron spins would then yield the spin value of the neutron.

$$n \;\rightarrow\; p^+ \;+\; e^-$$
$$\text{spin:} \quad (\pm\tfrac{1}{2}) \neq (\pm\tfrac{1}{2}) + (\pm\tfrac{1}{2})$$

Thus spin is clearly not conserved in such a process. Linear momentum is violated by virtue of the observation that the daughter nucleus does not recoil in directions opposite to the electron's motion.

Observations on beta decay thus confronted the physics community with a most serious dilemma. The choice became either to reject these conservation laws as being no longer of universal validity or to resolve the matter in a way so as to be consistent with them. The former alternative, however, had implications that reached far beyond the phenomena of beta decay. Space, for example, would no longer be either homogeneous or isotropic, nor would time be intrinsically homogeneous.

The Neutrino Hypothesis

In 1930 W. Pauli (1900–1958) suggested an alternative to the abandonment of the conservation laws. He proposed that there is yet another particle involved in the beta decay process. The neutron decay, for example, would then produce this new particle in addition to a proton and an electron. As for its properties, this hypothetical particle could not carry electric charge, because charge stood conserved without it, nor could it have an appreciable rest mass, because those few electrons carrying the maximum available energy were already conserving energy without this hypothetical particle. Even though it could have neither charge nor mass, the hypothetical particle still had to possess a spin of $\frac{1}{2}$, for otherwise its existence would not help conserve spin.

This new particle is somewhat reminiscent of the photon, which also has neither charge nor mass. It, like the photon, is able to carry kinetic energy and momentum, so that the missing energy and momentum in beta decay can be attributed to these hypothetical particles. Unlike the photon, however, these particles must be very elusive; in fact, so elusive that they must trigger no detector.

Pauli's hypothesis of the existence of an elusive particle in beta decay thus achieved the restoration of universal validity to the conservation laws of momentum, energy, and spin. Pauli's hypothesis was placed on a still firmer foundation when it was successfully incorporated by E. Fermi (1901–1954) into a theory of beta decay. Fermi christened the undetected particle the neutrino, which means the "little neutral one" in Italian.

The Fermi Theory of Beta Decay

New ideas are often proposed in analogy with old ideas. Fermi's formulation of a theory of beta decay was no exception. He envisioned the emission of beta particles as being conceptually similar to the emission of a photon from an atom. The emission of a photon, in the quantum theory of radiation developed initially by Dirac and later by Fermi himself, is essentially a creation process. The creation of a photon by an atom takes place at the moment of a quantum transition in the atom. In formulating his theory of beta decay, Fermi proposed that beta decay also be viewed as a creation process. That is, the transformation of a nucleon from one charge state to another charge state was to be achieved by the creation of a pair of particles, namely, a beta particle and a neutrino. The beta particle and the neutrino, created simultaneously at the instant of nucleon transformation, are then described quantum mechanically by Dirac's relativistic wave equation. Because Dirac's wave equation admits the existence of antimatter, Fermi's theory of beta decay describes the emission of particles as well as antiparticles.

In the Fermi theory, the spontaneous emission of an electron from a radioactive nucleus arises from the transformation of a neutron into a proton with the creation of an electron–antineutrino pair.

$$n \rightarrow p^+ + e^- + \bar{\nu}$$

where the bar above the neutrino, $\bar{\nu}$, denotes an antineutrino. The emission of a positron occurs when a proton inside the nucleus is transformed into a neutron by the creation of a positron–neutrino pair.

$$p^+ \rightarrow n + e^+ + \nu$$

Radioactivity

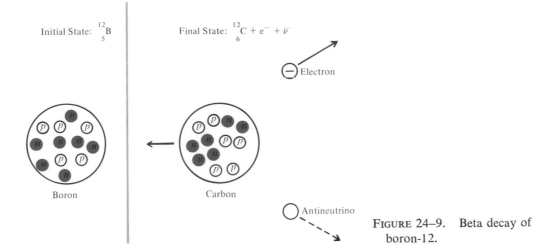

Initial State: $^{12}_{5}\text{B}$ Final State: $^{12}_{6}\text{C} + e^- + \bar{\nu}$

Electron

Boron   Carbon

Antineutrino

FIGURE 24–9.   Beta decay of boron-12.

Of course, it must be realized that a free proton outside the nucleus is forbidden by the conservation of energy to transform into a neutron, because the neutron has a higher rest energy than the proton. Inside the nucleus, however, the energy of the nucleus itself is involved in such a transformation, and a proton can indeed turn into a neutron.

In certain nuclei a proton is transformed into a neutron without the emission of a beta particle. Such a transformation is also described by the Fermi theory, in which the proton captures an innermost atomic electron with a simultaneous creation of neutrino.

$$e^- + p^+ \rightarrow n + \nu$$

Notice that, in all three types of nucleon transformation in beta decay, a pair of nucleons is involved as well as a pair of a beta particle and a neutrino. The beta particle and the neutrino are classified as belonging to the electron family. The neutron and the proton, according to a similar classification scheme, belong to the so-called baryon family, which includes the most massive particles in nature. The proton is the lightest member of the baryon family. The occurrence of these particles in pairs in beta decay is suggestive of a conservation law. It is, indeed, found that the total family number is conserved not only in beta decay but in all physical processes. The family number assigns a positive unity to a particle member of a family, and a negative unity to an antiparticle member. To a particle not belonging to a particular family a zero is assigned in such a counting scheme. The conservation of the baryon family number and of the electron family number may be illustrated as follows.

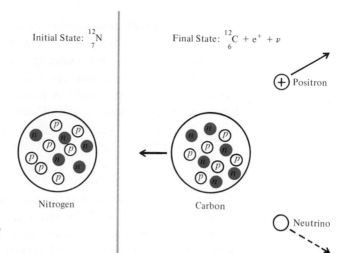

Initial State: $^{12}_{7}\text{N}$

Final State: $^{12}_{6}\text{C} + e^+ + \nu$

(+) Positron

Nitrogen

Carbon

○ Neutrino

FIGURE 24–10.   Beta decay of
nitrogen-12.

$$n \;\rightarrow\; p^+ \;+\; e^- \;+\; \bar{\nu}$$

electron family number: $0 \;=\; 0 \;+\; (+1) + (-1)$

baryon number: $(+1) + (+1) +\; 0 \;+\; 0$

TABLE 24–1
BARYONS

Particle	Spin	Mass (MeV)	Lifetime (sec)
p proton	$\frac{1}{2}$	938.256	stable
n neutron	$\frac{1}{2}$	939.550	1.01×10^3
Λ lambda particle	$\frac{1}{2}$	1115.58	2.51×10^{-10}
Σ^+ sigma particles	$\frac{1}{2}$	1189.47	0.81×10^{-10}
Σ^0	$\frac{1}{2}$	1192.56	$<1.0 \times 10^{-14}$
Σ^-	$\frac{1}{2}$	1197.44	1.65×10^{-10}
Ξ^0 cascade particles	$\frac{1}{2}$	1314.7	3.0×10^{-10}
Ξ^-	$\frac{1}{2}$	1321.2	1.74×10^{-10}
Ω^- omega minus	$\frac{3}{2}$	1674	1.5×10^{-10}

Radioactivity

These, along with other conservation laws, help maintain stability and order in the universe. Conservation laws are essentially laws of prohibition: no physical process violating a conservation law can occur in nature. Baryon family conservation, for example, would not allow a proton to move out of the family. A proton, therefore, cannot decay into a lighter particle, because it is itself the lightest member of the baryon family. The observed stability of the proton is a clear demonstration of the validity of baryon family conservation.

The conservation of the electron family number is not explicitly assumed by the Fermi theory of beta decay. The simultaneous creation of a beta–neutrino pair in beta decay, however, establishes the neutrinos as bona fide members of the electron family. Once neutrinos are so established, electron family conservation implies that an electron must be created with an antineutrino, and a positron must be created with a neutrino.

The processes of electron emission, positron emission, and electron capture are all beautifully described by the Fermi theory. The theoretical predictions are in good agreement with experimental results. The theory, in producing a continuous energy spectrum for the beta particles, attributes the missing energy to the undetected neutrinos and thus puts Pauli's neutrino hypothesis on a firm theoretical basis.

Fermi also showed that the forces responsible for the creation of a beta–neutrino pair are much weaker than the nuclear force or the electromagnetic force. In fact, it is so much weaker than the photon-creating electromagnetic force that the creation of an electron–antineutrino pair takes anywhere from a few seconds to centuries, whereas an atomic transition via photon emission requires typically about 10^{-8} seconds. It is this inherent weakness of the interaction force that makes the neutrino so breathtakingly elusive.

QUESTIONS

24–1. What is the binding energy per nucleon of deuteron?

24–2. Explain why a very heavy nucleus, such as uranium, splitting into two nuclei of intermediate mass, such as krypton and barium, would be accompanied by a release of energy.

24–3. A portion of the carbon in plants is radioactive carbon-14.

When a plant dies, no additional carbon-14 is taken in from the atmosphere, so that the ratio of carbon-14 to carbon-12 in the plant decreases with the passage of time. Some interesting archaeological events have been dated by means of the carbon-14 dating technique.

(a) Explain why radioactive decay can provide a very convenient way of measuring the passage of time.

(b) Why is the carbon-14 dating technique particularly suited for archaeological dating measurements?

24–4. List the properties of the neutrino and explain how each assigned property helps make the conservation laws of energy, momentum, and spin valid in beta decay.

24–5. The neutrino and the electron, along with their antimatter counterparts, are the only members of the electron family. Explain why the electron does not decay into a neutrino and a photon, when such could clearly conserve energy and the electron family number.

24–6. The neutrino hypothesis is theoretically very attractive.

(a) Discuss the theoretical support for the hypothesis.

(b) Could the neutrino be regarded as a physical entity solely on theoretical grounds? Explain your position.

24–7. Comment on the statement that conservation laws are laws of prohibition.

SUGGESTIONS FOR FURTHER READING

GAMOW, G. *The Atom and Its Nucleus*. Englewood Cliffs, N.J.: Prentice-Hall, Inc., 1961.

LIBBY, W. F. *Radiocarbon Dating*. Chicago: University of Chicago Press, 1965.

MARION, J. B. *A Universe of Physics*. New York: John Wiley and Sons, Inc., 1970. (Chapter 7.)

SEGRÈ, E. *Enrico Fermi, Physicist*. Chicago: University of Chicago Press, 1970.

Radioactivity_____

25
The Neutrino

The neutrino is the most elusive physical entity known to man. This is by no means to imply that the neutrino is devoid of physical information. On the contrary, the neutrino has revealed certain intricacies about the universe that no other particle is capable of doing, and may in time provide a clue to the very nature of the universe.

Detection of the Neutrino

Despite the impressive success of the Fermi theory and the conceptual soundness of the neutrino hypothesis, there was still a strong feeling in some quarters that the neutrino was not a real physical entity but a clever myth designed to save the "sacred" conservation laws. As long as the neutrino remained undetected, such an attitude of skepticism was not entirely unwarranted.

Any successful detection of the neutrino had to involve a strong source of neutrinos, because the number of neutrinos absorbed by matter would be incredibly small because of its extremely weak interaction with matter. The first convincing evidence of the existence of the neutrino was provided by an experiment of C. L. Cowan, Jr., and F. Reines in 1956. They employed as a source of antineutrinos the world's then-largest nuclear reactor at the A.E.C.'s Savannah River Plant in South Carolina. The flux of the antineutrinos emerging from the reactor was known to be some 5×10^{13} particles per square centimeter per second. Two hundred liters of water served as a target for the antineutrinos, because antineutrinos would react with the hydrogen nuclei of water. Calculation showed that reaction events would occur about once every twenty minutes.

The antineutrino reaction with the proton proceeds as:

$$\bar{v} + p^+ \rightarrow n + e^+$$

The absorption of an antineutrino by a proton is accompanied by the creation of a neutron and a positron. The experiment consisted in pinpointing such events. It was known that the positron would be annihilated by an electron in the water within 10^{-9} seconds, thereby emitting two gamma rays of equal energy in opposite directions. As a way of detecting the neutron, the water target was seasoned with cadmium chloride. Cadmium is a good absorber of neutrons and would indicate its absorption of a neutron by emitting several gamma rays. The neutrons produced from antineutrino reactions would suffer absorption by cadmium in about 10^{-6} seconds. An antineutrino reaction with a proton would thus be evidenced by two photons of equal energy from positron annihilation, followed some 10^{-6} seconds later by gamma rays from the cadmium excited by neutron capture. Such a sequence of events was not only uniquely attributable to the absorption of an antineutrino by a proton, but the number of events thus singled out also agreed with the prediction of the Fermi theory. There was, therefore, no doubt that reactions caused directly by antineutrinos had, indeed, been detected.

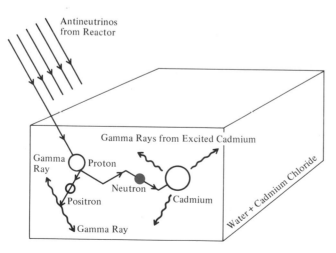

FIGURE 25–1. Schematic diagram of the antineutrino-detection experiment by Reines and Cowan.

Physics: The Fabric of Reality

Are the Neutrino and the Antineutrino Identical?

The Fermi theory does not determine whether the neutrino is distinguishable from the antineutrino. But the question is of great significance in that their identical character would violate the conservation of the electron family number. An answer was sought in a reactor experiment by R. Davis, Jr., about the time of the Cowan–Reines experiment. Davis subjected chlorine to a flux of antineutrinos from a reactor and sought argon nuclei and electrons as reaction products.

$$\bar{\nu} + {}^{37}\text{Cl} \rightarrow {}^{37}\text{Ar} + e^-$$
$$\text{electron family number:} \quad (-1) + 0 \neq 0 + (+1)$$

An observation of such a reaction would clearly violate electron family conservation. Such a reaction, however, was found to occur at least a thousand times less frequently that the antineutrino reactions observed by Cowan and Reines, which did conserve the electron-family number. The negative result of the Davis experiment was interpreted as an evidence of prohibition by the conservation of the electron-family number. The experiment further suggested that the neutrino is fundamentally different from the antineutrino.

How is the neutrino different from the antineutrino? They are, to be sure, a particle and an antiparticle, respectively, but there are a number of particles in the universe which are identical to their antiparticles. So the difference between the neutrino and the antineutrino must be much more profound than their matter–antimatter distinction.

The Neutrino's Violation of the Mirror Symmetry

When one looks in the mirror, one does not ordinarily wonder if the image in the mirror is really a likeness of oneself. But what would the physical world be like if the photograph of a mirror image were fundamentally different from that of the original object?

Physicists have always believed that nature possessed a mirror symmetry—as least, until some two decades ago. If nature is mirror symmetric, the image of a physical process would be indistinguishable from the process itself. Mirror symmetry is equivalent to the symmetry between right and left, as a right-handed process would appear to be a left-handed process in its mirror image. A mirror-symmetric universe, then, would not distinguish the right hand from the left hand. There would be no intrinsically right-handed or left-handed process in such a universe.

More precisely, the laws of physics describing a mirror-symmetric universe would themselves possess a mirror symmetry. This is to say that a physical law describing a particular phenomenon would also describe its mirror-image phenomenon with equal validity. Thus there would exist no experiment in such a world that would distinguish a phenomenon from its mirror image.

When they believed nature to possess a right and left symmetry, physicists had a good reason. The universe is governed by four fundamental interactions: the gravitational interaction, the electromagnetic interaction, the "strong" interaction, and the "weak" interaction. The gravitational interaction holds together the universe at large. The electromagnetic interaction governs the atomic phenomena, binding electrons to the nucleus and binding atoms to form molecules and solids. The "strong" interaction holds the nucleus together. The fourth interaction, referred to as the "weak" interaction, describes beta-decay processes, and thus the world of the neutrino. The first three interactions had long been established to be mirror symmetric. To extend the mirror symmetry to the "weak" interaction thus seemed to be a natural thing to do, if for none other than aesthetic reasons. After all, is symmetry not a symbol of "perfection"?

A mirror-symmetric interaction is said to conserve parity, as the prin-

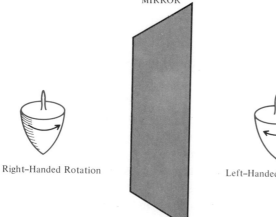

Right–Handed Rotation

MIRROR

Left–Handed Rotation

FIGURE 25–2. A top in a mirror-symmetric world would rotate with equal facility in both right-handed and left-handed directions.

Physics: The Fabric of Reality

ciple of right and left symmetry is also called the principle of conservation of parity. Parity can, by definition, assume two states: even or odd. Conservation of parity would then imply that a physical process would preserve its "evenness" or "oddness," no matter what changes it might undergo.

In the middle 1950's the physics community was confronted with a puzzling situation. A particle seemed to posses both an even and an odd parity. It decayed either into two pions or into three pions. When it decayed into two pions, it was called a θ-particle and was assigned an even parity, because two pions together have an even parity. When it decayed into three pions, it was called a τ-particle and was assigned an odd parity, because three pions together possess an odd parity. Attempts to establish the τ- and θ-particles as different particles, however, all failed. On the contrary, it became increasingly clear that the τ- and θ-particles were, indeed, identical. This was the famous "τ–θ puzzle."

The decay of these particles is described by the weak interaction theory of Fermi. As long as the weak interaction is assumed to be mirror symmetric, the situation remained a puzzle. But to admit a violation of parity conservation in weak interactions would also be puzzling. For such would admit the possibility that a process proceeding via a weak interaction would have no mirror-image counterpart in nature.

A symmetric sense of reality did apparently not dominate the thinking of the Chinese physicists, T. D. Lee and C. N. Yang. They saw in the "τ–θ puzzle" a suggestion of the possibility that nature may not be perfectly symmetrical, that nature may indeed distinguish between left and right. They pursued the idea by first searching the scientific literature to see if the conservation of parity, which had been firmly established for the strong, electromagnetic, and gravitational interactions, had also been tested for the weak interaction. Finding no experimental evidence for it, they proposed an experiment to test the possible violation of parity conservation in beta decay. On the theoretical side, they argued that violation of a right and left symmetry in nature would help explain a number of mysteries.

The suggestion of Lee and Yang in 1956 stirred the imagination of another Chinese physicist, C. S. Wu. The question of possible parity non-conservation held such urgency for her that Mme. Wu immediately planned an experiment. The experiment examined the beta decay of cobalt-60. The cobalt nucleus transforms into a nickel-60 nucleus by emitting an electron and an antineutrino.

The Neutrino ————————————————————————

$$^{60}\text{Co} \rightarrow {}^{60}\text{Ni} + e^- + \bar{\nu}$$

The cobalt nucleus has a spin, and thus acts like a little bar magnet. If parity is conserved, the electron would be emitted from the north pole of the cobalt magnet as readily as from the south pole. The reason is that the mirror image of an electron being emitted from the north pole of a nucleus is the same as an electron being emitted from the south pole of the nucleus. Because both situations would be equally probable according to parity conservation, the electron would have no preferential direction of emission.

Mme. Wu and her collaborators aligned the cobalt-60 nuclei so as to point all in the same direction by means of a strong external magnetic field. The desired alignment was achieved by reducing the temperature of the cobalt sample to 0.01° K. They then detected the electron emission, both from the "top" and the "bottom" of the cobalt-60 sample. The result was decisive. The electrons emerged from the "top" some 40 per cent more than from the "bottom." The electrons clearly emerged in a

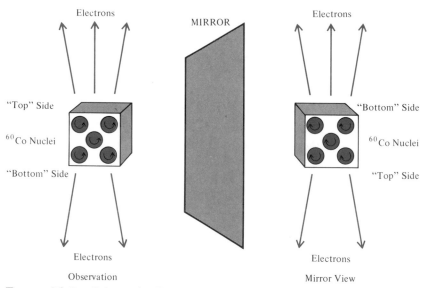

FIGURE 25–3. Schematic diagram of the cobalt-60 beta decay experiment by C. S. Wu. It is observed that electrons are preferentially emitted from the "top" side of the aligned cobalt-60 nuclei, with the result that the mirror image of the phenomenon has no counterpart in nature.

Physics: The Fabric of Reality

preferred direction. The mirror image of the radioactive cobalt nucleus, thus, did not exist in nature. Here was an experimental proof that parity is not conserved in beta decay.

The Right-Handed Antineutrino

The violation of parity in beta decay was pinpointed to the neutrino. Lee and Yang argued that the parity violation in the cobalt-60 decay occurs because the antineutrino possesses an intrinsic right-handed rotation. A right-handed antineutrino spins in such a way that its motion is parallel to the axis of spin and its rational sense forms a right-handed screw. As a massless particle, the antineutrino always travels at the speed of light. The right-handedness of the antineutrino, then, is a property invariant to all observers. It may thus be regarded as an intrinsic property. But the mirror image of a right-handed antineutrino would be a left-handed antineutrino, which does not exist. The antineutrino, therefore, possesses no mirror image. Mirror symmetry does not exist in the world of the antineutrino.

The preferential direction of the emission of the electrons from cobalt nuclei may now be explained in terms of the antineutrino's right-handedness. The spin of the cobalt nucleus is 5 in units of $h/2\pi$, and the spin of the daughter nucleus, nickel-60, is 4 in the same units. This spin, furthermore, is a right-handed rotation. Thus a right-handed antineutrino must emerge from the "top" of the spinning nucleus in order to conserve the rotational handedness, i.e., to conserve spin. But it is a well-known property of beta decay that the electrons prefer to accompany the antineutrinos in generally the same direction, with a rotational handedness similar to that of the antineutrino. The electrons would thus prefer to emerge from the "top" of the cobalt sample.

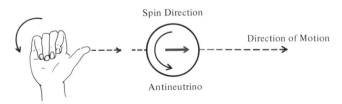

FIGURE 25–4.   The right-handed antineutrino.

The Neutrino

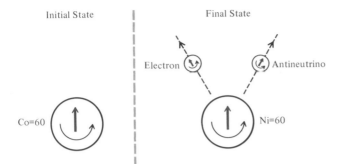

Initial State   Final State

Electron

Antineutrino

Co=60

Ni=60

FIGURE 25–5. The assymetric electron emission of cobalt-60, as explained by Lee and Yang.

The Left-Handed Neutrino

The violation of parity conservation in weak interactions implies that the neutrino must also exhibit a handedness. The handedness of the neutrino can be established from an observation of the cobalt-58 decay. The cobalt-58 nucleus emits a positron and a neutrino, thereby becoming iron.

$$^{58}\text{Co} \rightarrow {}^{58}\text{Fe} + e^+ + \nu$$

When the cobalt nuclei are aligned to exhibit a right-handed rotation, the positrons are predominantly emitted from the "bottom," and so must the neutrinos be. As the neutrinos move downward with respect to the spin of the nucleus, they must still exhibit a rotational sense similar to the right-handed rotation of the nucleus. The spin of the neutrino must then be antiparallel to the direction of its motion. (Recall that the spin direction of an antineutrino is parallel to the direction of its motion.) The neutrino is thus said to exhibit a left-handedness.

In short, the neutrino is fundamentally different from the antineutrino, because the universe deviates slightly from a perfect right and left symmetry. Looking in the mirror, a neutrino would see an image that has no counterpart in nature. Or, put another way, the image world of the neutrino distorts the mirror symmetry in nature, thus rendering the universe asymmetrical—perhaps a little less than "perfect."

Two Types of Neutrinos

Consider now the beta decays of the neutron and the pion. In the case of the neutron an electron is emitted, accompanied by an antineutrino. The pion, on the other hand, decays into a muon and a neutrino. The

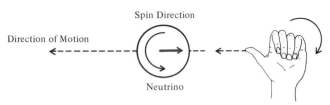

Spin Direction

Direction of Motion

Neutrino

FIGURE 25–6.   The left-handed neutrino.

question then arises: Is the neutrino created with an electron identical to the neutrino created with a muon? As no existing theory provides an answer to that question, the issue must be settled by experiment.

The question is important, because the muon is identical to the electron in all of its interactions except for a difference in mass. The muon is about 200 times as massive as the electron. The origin of the muon's anomalously large mass is a great mystery. It is also puzzling that nature has seen fit to admit it into her scheme of things, when the muon seems completely redundant as no more than a merely "heavy" electron.

The conceptual basis for a two-neutrino experiment is simple. The theory of weak interactions predicts that if a neutrino is absorbed by a neutron, a proton and an electron would be produced.

$$\nu + n \rightarrow p^+ + e^-$$

If the energy of the neutrino is increased so as to provide enough energy to produce a muon, the reaction would also produce a proton and a muon.

$$\nu + n \rightarrow p^+ + \mu^-$$

Thus if the muon-type neutrino is identical to the electron-type neutrino, the absorption of a high-energy neutrino by a neutron would result in the creation of equal numbers of electrons and muons.

An experiment, performed by L. Lederman, M. Schwartz, and J. Steinberger in 1962, consisted in bombarding a target with a beam of neutrinos from the decay of a pion beam. An energetic pion beam can readily be produced in a powerful accelerator. As these pions decay in flight, muons and muon-type neutrinos are produced. The muon-type neutrinos move predominantly in the forward direction, thus forming a high-energy neutrino beam. Upon bombarding a target, such a neutrino beam would then produce equal numbers of electrons and muons if these neutrinos were identical to the electron-type neutrinos.

The experimental result was that only muons and no electrons were produced by the neutrino beam. The observation was thus interpreted as establishing two types of neutrinos. The neutrino created with a muon,

The Neutrino

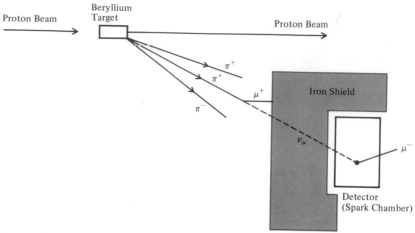

FIGURE 25–7.   Schematic diagram of the production of a muon-type neutrino beam. About 3×10^{14} neutrinos passed through the spark chamber, and twenty-nine significant events were recorded. (Adapted from L. M. Lederman.)

as in the decay of a pion, was simply not capable of producing an electron. Similarly, a neutrino produced with an electron would not be able to create a muon, although an electron-type neutrino beam energetic enough to produce muons is not available. The muon-type and the electron-type neutrinos must therefore be viewed as fundamentally different. No physical difference, however, is apparent between them. They both seem to have identical properties: no mass, no charge, and spin ½.

When the muon was first discovered, the muon was thought to belong to the electron (or lepton) family. However, the difference between the muon-type and the electron-type neutrino now separates the muon family from the electron family, and each family number is conserved separately. A muon, decaying into an electron and two neutrinos, produces actually two different types of neutrinos. The family number is then assigned to the neutrinos by invoking the conservation of family numbers.

$$\mu^- \rightarrow e^- + \bar{\nu}_e + \nu_\mu$$

muon number: $(+1) = 0 + 0 + (+1)$
electron number: $0 = (+1) + (-1) + 0$

Conclusion

The neutrino has, indeed, been a source of delight and fascination. It is a tribute to the creative power of the human mind that, from a modest beginning in 1930, the neutrino has blossomed into a shining thread in the fabric of reality. It is, moreover, a thread linking the entire universe. As man ponders the mysteries of the universe, it is an awesome and exciting thought that the neutrino may, as some now think, hold a vital and ultimate clue to the very nature and the origin of the universe.

QUESTIONS

25–1. What part did theory play in the experimental observation of the antineutrino?

25–2. The Davis experiment can be interpreted as suggesting a fundamental difference between the neutrino and the antineutrino. Explain.

25–3. If an electron appears right-handed to one observer, would the same electron ever appear left-handed to another observer? Explain.

Answer: Yes

25–4. Explain how the right-handedness of the antineutrino is established.

25–5. Explain why the right-handedness of the antineutrino violates the principle of parity conservation.

25–6. What is the fundamental assumption in the two-neutrino experiment?

25–7. Explain why muon decay into an electron and a photon is forbidden.

SUGGESTIONS FOR FURTHER READING

ALFVÉN, H. *Worlds-Antiworlds*. San Francisco: W. H. Freeman and Company, 1966.

FORD, K. W. *Basic Physics*. Waltham, Mass.: Blaisdell Publishing Company, 1968. (Chapters 26 and 27.)

The Neutrino

GAMOW, G. *The Creation of the Universe*. New York: Viking Press, 1952.

GARDNER, M. *The Ambidextrous Universe*. New York: Basic Books, 1964.

LEDERMAN, L. M. "The Two-Neutrino Experiment," *Scientific American* (March 1963).

MORRISON, P. "The Overthrow of Parity," *Scientific American* (April 1957).

REINES, F., and C. L. COWAN. "Neutrino Physics," *Physics Today* (August 1957), pp. 12–18.

TREIMAN, S. B. "The Weak Interactions," *Scientific American* (March 1961).

Appendix

Nobel Prizes in Physics

1901	Wilhelm Konrad Röntgen *(Germany)*	Discovery of X-rays.
1902	Hendrik Antoon Lorentz *(Netherlands)* Pieter Zeeman *(Netherlands)*	Influence of magnetism on electro-magnetic radiation.
1903	A. Henri Becquerel *(France)* Pierre Curie *(France)* Marie Curie *(France)*	Discovery of radioactive elements.
1904	John Strutt (Lord Rayleigh) *(England)*	Discovery of argon.
1905	Philipp Lenard *(Germany)*	Research on cathode rays.
1906	Sir Joseph John Thomson *(England)*	Conduction of electricity through gases.
1907	Albert A. Michelson *(U.S.A.)*	Spectroscopic and metrological investigations.
1908	Gabriel Lippman *(France)*	Color photography.
1909	Guglielmo Marconi *(Italy)* Karl Ferdinand Braun *(Germany)*	Development of wireless telegraphy.

1910	Johannes Diderik van der Waals *(Netherlands)*	Equations of state of gases and fluids.
1911	Wilhelm Wien *(Germany)*	Laws of heat radiation.
1912	Nils Gustaf Dalén *(Sweden)*	Coast lighting.
1913	Heike Kamerlingh-Onnes *(Netherlands)*	Properties of matter at low temperatures; production of liquid helium.
1914	Max von Laue *(Germany)*	Diffraction of X-rays in crystals.
1915	Sir William Henry Bragg *(England)* William Lawrence Bragg *(England)*	Study of crystal structure by means of X-rays.
1916	No award	
1917	Charles Glover Barkla *(England)*	Discovery of the characteristic X-ray spectra of elements.
1918	Max Planck *(Germany)*	Quantum theory of radiation.
1919	Johannes Stark *(Germany)*	Discovery of the Doppler effect in positive ion beams and the splitting of spectral lines in an electric field.
1920	Charles Edouard Guillaume *(Switzerland)*	Discovery of anomalies in nickel–steel alloys.
1921	Albert Einstein *(Germany)*	The theory of relativity and the photoelectric effect.
1922	Niels Bohr *(Denmark)*	Quantum theory of the atom.
1923	Robert Andrews Millikan *(U.S.A.)*	Measurement of the elementary electric charge and the photoelectric effect.
1924	Karl Manne Siegbahn *(Sweden)*	Discoveries in the area of X-ray spectra.
1925	James Franck *(Germany)* Gustav Hertz *(Germany)*	Laws governing collisions between electrons and atoms.
1926	Jean Perrin *(France)*	Discovery of the equilibrium of sedimentation.

Physics: The Fabric of Reality⸺⸺⸺⸺⸺⸺⸺⸺⸺⸺

1927	Arthur H. Compton *(U.S.A.)*	Explanation of the scattering of X-rays by electrons and atoms.
	Charles T. R. Wilson *(England)*	Cloud chamber for particle detection.
1928	Sir Owen Williams Richardson *(England)*	Discovery of the law of thermionic emission of electrons.
1929	Prince Louis-Victor de Broglie *(France)*	Wave nature of electrons.
1930	Sir Chandrasekhara Raman *(India)*	Diffusion and scattering of light by molecules.
1931	No award	
1932	Werner Heisenberg *(Germany)*	Creation of quantum mechanics.
1933	Paul Adrien Maurice Dirac *(England)*	Development of relativistic quantum mechanics.
	Erwin Schrödinger *(Austria)*	Creation of quantum mechanics.
1934	No award	
1935	James Chadwick *(England)*	Discovery of the neutron.
1936	Victor F. Hess *(Austria)*	Discovery of cosmic radiation.
	Carl David Anderson *(U.S.A.)*	Discovery of the positron.
1937	Cliton Joseph Davisson *(U.S.A.)*	Discovery of diffraction of electrons by crystals.
	George P. Thomson *(England)*	
1938	Enrico Fermi *(Italy)*	Artificial radioactivity induced by slow neutrons.
1939	Ernest O. Lawrence *(U.S.A.)*	Invention of the cyclotron.
1940	No award	
1941	No award	
1942	No award	
1943	Otto Stern *(Germany)*	Detection of magnetic moment of protons.
1944	Isidor Isaac Rabi *(U.S.A.)*	Magnetic moments by beam methods.
1945	Wolfgang Pauli *(Austria)*	The exclusion principle.

Appendix

1946	Percy Williams Bridgman (U.S.A.)	High-pressure physics.
1947	Sir Edward Appleton (England)	Discovery of ionospheric reflection of radio waves.
1948	Patrick Maynard Stuart Blackett (England)	Discoveries in cosmic radiation.
1949	Hideki Yukawa (Japan)	Theoretical prediction of the meson.
1950	Cecil Frank Powell (England)	Photographic method of studying atomic nuclei; discovery of the charged pion.
1951	Sir John Douglas Cockcroft (England) Ernest Thomas Sinton Walton (Ireland)	Transmutation of atomic nuclei by accelerated atomic particles.
1952	Felix Bloch (U.S.A.) Edward Mills Purcell (U.S.A.)	Measurement of magnetic fields in atomic nuclei.
1953	Fritz Zernike (Netherlands)	Development of phase contrast microscopy.
1954	Max Born (Germany)	Contributions to quantum mechanics.
	Walther Bothe (Germany)	Studies of cosmic radiation by use of the coincidence method.
1955	Willis E. Lamb, Jr. (U.S.A.) Polykarp Kusch (U.S.A.)	Atomic measurements.
1956	John Bardeen (U.S.A.) Walter H. Brattain (U.S.A.) William B. Schockley (U.S.A.)	Invention and development of the transistor.
1957	Chen Ning Yang (China, U.S.A.) Tsung Dau Lee (China, U.S.A.)	Overthrow of principle of conservation of parity.
1958	Pavel A. Cerenkov (U.S.S.R.) Ila M. Frank (U.S.S.R.) Igor Y. Tamm (U.S.S.R.)	Interpretation of radiation effects and development of a cosmic-ray counter.
1959	Owen Chamberlain (U.S.A.) Emilio Gino Segrè (U.S.A.)	Discovery of the antiproton.
1960	Donald A. Glaser (U.S.A.)	Invention of bubble chamber.

Physics: The Fabric of Reality———————————————

1961	Robert L. Hofstadter *(U.S.A)*	Electromagnetic structure of nucleons from high-energy electron scattering.
	Rudolph L. Mössbauer *(Germany)*	Discovery of recoilless resonance absorption of gamma rays in nuclei.
1962	Lev D. Landau *(U.S.S.R.)*	Theories of condensed matter.
1963	Eugene P. Wigner *(U.S.A.)* Marie Goeppert-Mayer *(U.S.A.)* J. Hans D. Jensen *(Germany)*	Nuclear shell structure.
1964	Charles H. Townes *(U.S.A.)* Nikolai G. Basov *(U.S.S.R.)* Alexander M. Prokhorov *(U.S.S.R.)*	Amplification by laser–maser devices.
1965	Richard P. Feynman *(U.S.A.)* Julian S. Schwinger *(U.S.A.)* Sin-Itiro Tomonaga *(Japan)*	Quantum electrodynamics.
1966	Alfred Kastler *(France)*	Atomic energy levels.
1967	Hans A. Bethe *(Germany)*	Nuclear theory and stellar energy production.
1968	Luis W. Alvarez *(U.S.A.)*	Elementary particles.
1969	Murray Gell-Mann *(U.S.A.)*	Theory of elementary particles.
1970	Hannes Alfvén *(Sweden)*	Astrophysics
1971	Dennis Gabor *(U.S.A.)*	Holography
1972	John Bardeen *(U.S.A.)* Leon Cooper *(U.S.A.)* John Schrieffer *(U.S.A.)*	Theory of superconductivity.
1973	Leo Isaki *(Japan)* Ivar Giaever *(U.S.A.)* Brian Josephson *(England)*	Discoveries in semiconductor physics. Prediction of "Josephson effects" in superconductivity.

Index

Einstein, Albert (1879–1955), 24, 26, 30, 34, 37, 57, 72, 73, 74, 107, 108, 109, 110, 112, 114, 116, 117, 118
Electric charge: *see* Charge
Electric current, 145
Electric field, 131–142
    atmospheric, 140, 142
    changing, 159
    definition of, 135
    lines, 136, 137
Electric force, 133–134
Electric potential, 137
    atmospheric, 138–142
Electromagnetic energy, 165, 166
Electromagnetic field, 145, 148, 160, 161, 162, 163
    quantized, 249
Electromagnetic induction, 159, 160
Electromagnetic momentum, 165, 166
Electromagnetic radiation, 165, 166
    mass of, 166
Electromagnetic theory, 10, 15, 24, 26, 160, 169, 182, 197, 198, 240
Electromagnetic waves, 10, 163–165
Electron:
    charge, 132
    diffraction, 220–221, 253
    discovery, 193
    in a linear box, 234–235
    mass, 85
    spin, 268
    wave nature, 215–222
Electron family, 270
    conservation, 270, 277
Electron volt (eV), 138
Electrostatic force: *see* Coulomb's law
Ellipse, 122, 123
Emission of radiation: *see* Induced emission; spontaneous emission
Energy:
    conservation of, 72, 87, 163
    kinetic, 73–74, 80

Energy: [*cont.*]
    of light, 69
    and mass equivalence, 69–75, 79, 85, 92
    relativistic, 79–80
Energy states, negative, 86–89
Energy-level diagrams, 85
    of classical oscillator, 182
    of hydrogen atom, 199
    of nuclei, 261–262
    of quantum oscillator, 184
    of relativistic electron, 85–87
Energy–momentum relation, 80–81
Equivalence of mass and energy: *see* Energy
Ether, 9–12
Ether-drag hypothesis, 24
Event, 98
Exchange force, 249
Excited state, 201–202
Exclusion principle, 89, 210–212

Faraday, Michael (1791–1867), 10
Fermi, Enrico (1901 1954), 269, 272
Fermi theory of beta decay, 269–272
Feynman, Richard P., 12, 99, 100, 101, 102, 103
Feynman diagrams, 102, 103, 104, 250, 251, 252, 253
Fission, 75, 261, 262
FitzGerald, George F. (1851–1901), 24
Force, 121, 123, 124, 125, 248, 249, 250, 252
Ford, Kenneth W., 142
Frames of reference, 4–5
    absolute, 10, 12
    accelerated (or noninertial), 57, 107
    inertial, 24, 26
Frequency, 164
Frisch, D. H., 35, 39, 41
Fundamental interactions, 278

Fusion, 260

Galilean boat, 3, 5, 7
Galilean relativity, 6–7, 8
Galileo, Galilei (1564–1642), 2, 3,
 6, 7, 9, 109, 119, 120
Gamma decay, 262–263
Gamma rays, 263
Gamow, George, 90
Geiger, Hans, 195
General principle of relativity,
 107–108
Geocentric cosmology, 2
Germer, Lester H., 220, 221
Graph, space–time, 98
Gravitation, 107–116, 123–127
Gravitational constant, 126
Gravitational force, 125, 127
Gravitational interaction: see
 Interactions
Gravitational potential, 137
Gravitational red shift, 115
Gravity, 110, 112–114, 127
Ground state, 201

h, Planck's constant:
 numerical value of, 189
Half-life, 35, 262
Heisenberg, Werner K., 224, 234
 uncertainty principle, see
 Uncertainty principle
Helium nucleus: see Alpha particle
Hertz, Heinrich R. (1857–1894), 15,
 185
Hofstadter, Robert:
 experiment on nucleon charge
 structure, 253, 254, 255
Hydrogen atom, 191–202
 energy levels of, 199
 Bohr model of, 198–202
 ionization energy, 202
 radial probability densities,
 238–239
 Schrödinger theory of, 236–240
 size, 199, 200

Incandescent radiation, 181–183
Induced emission of radiation, 242
Induction, 159, 160
Inertia, 73, 121
 law of, 7, 120, 123
Inertial frame of reference, 6–7
Inner shells, 212
Interactions:
 electromagnetic, 248, 249, 272,
 278
 gravitational, 248, 249, 278
 strong (or nuclear), 249, 253, 272,
 278
 weak, 272, 278
Interference of waves, 218
Interval: see Space interval; Time
 interval
Invariance, 7
 of the speed of light, 25–26
Invariant:
 energy–momentum, 96
 relativistic, 95
 space–time, 96–97
Ionization energy, 202

Kepler, Johannes (1571–1630), 122,
 123, 192
Kepler's laws:
 first, 122, 124
 second, 122
 third, 123, 124
Kinetic energy: see Energy

Laser beam, 26
Lederman, L., 283
Lee, Tsung Dao, 279
Length contraction, 24, 41–50
Lepton family, 284
Lifetime, 243
Light:
 deflection by gravity, 113–114
 electromagnetic nature of, 169–178
 energy of, 69, 186

Quantum electrodynamics, 100, 249
Quantum number:
    angular momentum, 209, 238
    magnetic, 209, 238
    principal, 199, 208, 238
    spin, 210, 268, 281
Quantum transition, 89, 200, 215

Radioactivity, 259–272
Rainbow, 178
Rebka, Glen A., Jr., 111
Red shift, 115
Reflection of light, 175–177
Reines, Frederick, 275
Relativity:
    of clock synchronization, 57–63
    of energy, 69–72
    of mass, 74–75
    of simultaneity, 66
    of space, 39–51
    of time, 29–37
Rotating frame of reference, 112
Rotation, 109
Rutherford, Ernest (1871–1937),
        195, 202

Scattering:
    of alpha particles, 195–196
    of light, 169–173
    of X rays, 188
Schrödinger, Erwin (1887–1961),
        229, 233, 234
    quantum theory, 229–243
Schwartz, M., 283
Simultaneity, 66
Size:
    of atoms, 247
    of nuclei, 247
    of proton, 255
Smith, J. H., 35, 39, 41
Sommerfeld, Arnold J. W.
        (1868-1951), 208
Space:
    absolute, 26

Space: [*cont.*]
    contraction; *see* Length contraction
    curvature, 111–112
    homogeneity, 268
    interval, 42, 98–99
    isotropy, 268
Space–time, 27, 116
Space–time graph: *see* Feynman
        diagrams; Minkowski graph
Space–time union, 98
Space travel, 53
Spectral lines, 191, 200
Spectrometer, 191
Speed: *see* Velocity
Speed of light:
    constancy of, 25–26
    numerical value of, 11
Spin quantum number, 210, 268, 281
Spontaneous emission of radiation:
    quantum description of, 242–243
Stanford Linear Accelerator, 49
Stationary state, 198–199, 235
Steinberger, J., 283
Stimulated emission: *see* Induced
        emission
Strong interaction: *see* Interactions
Sunset, 172–173
Superposition:
    of wave functions, 240–241
    of waves, 218
Symmetry: *see* Mirror symmetry

T-Θ puzzle, 279
Temperature, 181, 280
Thomson, Sir Joseph J. (1856–1940),
        193–194
Time:
    absolute, 26, 29–30
    dilation, 32–36, 39, 41
    homogeneity of, 268
    interval, 30, 98
    warp, 112–113
Transition probabilities, 243
Transitions: *see* Quantum transition
Transverse dimension, 43–44

Index _____

Period   I A

Key

	Name
Carbon ←	Name
₆ **C** ←	Symbol
12.011 ←	Atomic mass (weight)*
2,4 ←	Electron configuration

Atomic number →

Metalloids   Nonmetals

— Transition elements —

Period	I A	II A	III B	IV B	V B	VI B	VII B	VIII B	
1	Hydrogen ₁H 1.0080 1								
2	Lithium ₃Li 6.941 2,1	Beryllium ₄Be 9.01218 2,2							
3	Sodium ₁₁Na 22.9898 2,8,1	Magnesium ₁₂Mg 24.305 2,8,2							
4	Potassium ₁₉K 39.102 2,8,8,1	Calcium ₂₀Ca 40.08 2,8,8,2	Scandium ₂₁Sc 44.9559 2,8,9,2	Titanium ₂₂Ti 47.90 2,8,10,2	Vanadium ₂₃V 50.9414 2,8,11,2	Chromium ₂₄Cr 51.996 2,8,13,1	Manganese ₂₅Mn 54.9380 2,8,13,2	Iron ₂₆Fe 55.847 2,8,14,2	Cobalt ₂₇Co 58.9332 2,8,15,2
5	Rubidium ₃₇Rb 85.4678 2,8,18,8,1	Strontium ₃₈Sr 87.62 2,8,18,8,2	Yttrium ₃₉Y 88.9059 2,8,18,9,2	Zirconium ₄₀Zr 91.22 2,8,18,10,2	Niobium ₄₁Nb 92.9064 2,8,18,12,1	Molybdenum ₄₂Mo 95.94 2,8,18,13,1	Technetium ₄₃Tc 98.9062 2,8,18,14,1	Ruthenium ₄₄Ru 101.07 2,8,18,15,1	Rhodium ₄₅Rh 102.9055 2,8,18,16,1
6	Cesium ₅₅Cs 132.9055 -18,18,8,1	Barium ₅₆Ba 137.34 -18,18,8,2	Lanthanum ₅₇La 138.9055 -18,18,9,2	Hafnium ₇₂Hf 178.49 -18,32,10,2	Tantalum ₇₃Ta 180.9479 -18,32,11,2	Tungsten ₇₄W 183.85 -18,32,12,2	Rhenium ₇₅Re 186.2 -18,32,13,2	Osmium ₇₆Os 190.2 -18,32,14,2	Iridium ₇₇Ir 192.22 -18,32,15,2
7	Francium ₈₇Fr (223) -18,32,18,8,1	Radium ₈₈Ra 226.0254 -18,32,18,8,2	Actinium ₈₉Ac (227) -18,32,18,9,2	104	105				

Lanthanide series →

	Cerium	Praseodymium	Neodymium	Promethium	Samarium	Europium
	₅₈Ce 140.12 -18,20,8,2	₅₉Pr 140.9077 -18,21,8,2	₆₀Nd 144.24 -18,22,8,2	₆₁Pm (145) -18,23,8,2	₆₂Sm 150.4 -18,24,8,2	₆₃Eu 151.96 -18,25,8,2

Actinide series →

	Thorium	Protactinium	Uranium	Neptunium	Plutonium	Americium
	₉₀Th 232.0381 -18,32,18,10,2	₉₁Pa 231.0359 -18,32,20,9,2	₉₂U 238.029 -18,32,21,9,2	₉₃Np 237.0482 -18,32,22,9,2	₉₄Pu (242) -18,32,24,8,2	₉₅Am (243) -18,32,25,8

*Numbers in parentheses are mass numbers of most stable isotopes.